Keys to the Universe

By Don J. Pershing

authorHOUSE

1663 LIBERTY DRIVE, SUITE 200
BLOOMINGTON, INDIANA 47403
(800) 839-8640
www.authorhouse.com

First published by AuthorHouse 09/15/04

ISBN: 1-4184-4765-X (e)
ISBN: 1-4184-4766-8 (sc)

Printed in the United States of America
Bloomington, Indiana

This book is printed on acid-free paper.

Dedication

This book is dedicated to my wife, Ernestine. I would also like to dedicate it to my father Dr. Alvin V. Pershing and my brother Norman W. Pershing for their interest in astronomy. I thank the students and teachers of Sierra High School for the chance they gave me to try different experiments in the classroom – the students and teachers were great. Special thanks go to the typists of this book, Elizabeth Anne Albright & Archana Reddy, for their long hours and perseverance. Lastly, this book is dedicated to all those interested in astronomy.

Introduction

This book is a compilation of papers I have written regarding the study of the Universe. The main objective of this book is to introduce the concept of the Gravitational Structural Length. This represents the bending of space caused by a given mass. This concept is important because it builds on the theories of past astronomers and physicists such as Galileo and Sir Isaac Newton. It can be used as a shortcut and cuts down on the amount of time needed to solve problems. It is also very accurate, sometimes more accurate than the constants used by past scholars. Other scholars of Astronomy can use this material as a research tool.

Table of Contents

Galileo & The Start of Physics

Location: Pisa, Italy

- Galileo's experiments of free fall of different masses from the Tower of Pisa and the experiment of the pendulum in the cathedral of Pisa

Galileo was born in Pisa, Italy on February 15, 1564. Galileo discovered the law of falling spheres when dropped from the Tower of Pisa. He found that the Earth would always accelerate any two different size spheres of different weights dropped at the same time from a given height at a constant rate. Galileo dropped a ten-pound weight and a one-pound weight from the Tower of Pisa. Both weights were accelerated at a constant rate due to the Earth's constant gravitational pull on both weights. Both weights hit the ground at the same time.

Equation: $v_\downarrow = \sqrt{2Gh} = m/s$, free fall velocity of weight
Where G is the gravitational constant of the Earth
& h is the height from where the spheres were dropped

Galileo was only 20 years old at the time of this discovery.

Galileo's Gravity of the Earth at 45°

G_{Earth} at 45° = 9.80616 m/s^2

Using 1 meter = 3.28032 ft,

G_{Earth} at 45° = (9.80616 m/s^2)(3.28032 ft/m) = 32.1673427712 ft/s^2

Galileo's Calculation of Free Fall for any Mass

h = (½)(G_{Earth})(Time of Free Fall)^2
h = (½)(9.80616 m/s^2) (6.38675869831 s^2)
h= 200 m, distance of free fall of the dropped weights

Calculation of terminal velocity

$v_\downarrow = \sqrt{(2)\,(9.80616\ m/s^2)\,(200\ m)} = 62.629577677$ m/s

Note: The pull of gravity is constant for any mass. At 45° the gravity is 9.80616 m/s^2.

Galileo's Pendulum Concept

$$T = 2\pi \sqrt{l / G_{Earth}}$$

The period of a pendulum, $T = 2.00645942068$ seconds

Data:

$2\pi = 6.28318530718$

$G_{Earth} = 9.80616$ m/s^2, the acceleration of gravity at 45° of the Earth

$l = 1$ meter

1 m $/ 9.80616$ m/s$^2 = 0.101976716676$ s^2

$\sqrt{0.101976716676 \text{ s}^2} = 0.319337934915$ s

$T = (2\pi)(0.319337934915$ s$) = 2.00645942068$ s, period of a pendulum

A Few Astronomers and Physicists

1)   Claudius Ptolemy (1543)

In Ptolemy's book, <u>Mathematical Compositions,</u> he recorded star positions at Alexandria, Egypt. He became one of our first astronomers. Ptolemy catalogued over 1,000 stars and charted the irregular motion of the moon. His observations were made around 150 A.D. His work was called <u>The Almagest</u>. He followed up on Hipparchus' work on the stars.

2)   Hipparchus (527 B.C. – 514 B.C.)

Hipparchus did his observation on the island of Rhodes. He was also one of our first astronomers. He charted the positions of star positions. From the star positions, he deduced that the celestial stars rotate. This variation in rotation is called precession. The precession requires about 26,000 years for one cycle.

3)   Nicolaus Copernicus (1473 – 1543)

Copernicus was of Polish descent but he was a German astronomer. In his book, <u>Revolutions of the Celestial Spheres</u> (1543), Copernicus corrected Ptolemy's work. He studied the motions of the planets. He formed the foundation of our modern concept of the motion of the planets.

4)   Galileo Galilei (1564 – 1642)

Galileo's two experiments of the pendulum and masses of two different size spheres might be said to be the start of physics. It certainly showed that gravity of the earth existed and was a constant. Galileo built the telescope to study the sky. He found that Jupiter had satellites in 1610. He found that the moon was a sphere with mountains and valleys. He turned his telescope on the sun and discovered that the sun had spots on it. The studying of the sun eventually lead to his blindness.

Galileo published <u>Dialogues on the Two New Sciences</u> and <u>A Dialogue on the Two Principle Systems of the World</u>. The church did not like his ideas in his books about the planets moving. Galileo was put in jail for the rest of his life.

5)   Isaac Newton (1642 – 1727)

Newton is noted for his work on gravity. He worked on gravity during the Great Plague at Cambridge in 1665. The college was dismissed during this period. Newton built a reflecting telescope in 1668. In 1684 Newton began his work on the nature of gravity. He wrote the Treatise on Light and on the Optics and The Mathematical Principles of Natural Philosophy. In The Mathematical Principles of Natural Philosophy, on page 332 is his work on the precession of the equinoxes. He calculated a value of 50"/years. His work and discoveries of light is still a major work in modern physics today.

Precession of the Equinoxes

Precession of the North Pole Star

The precessional shift of the North Pole star rotates over a period of 26,000 years.

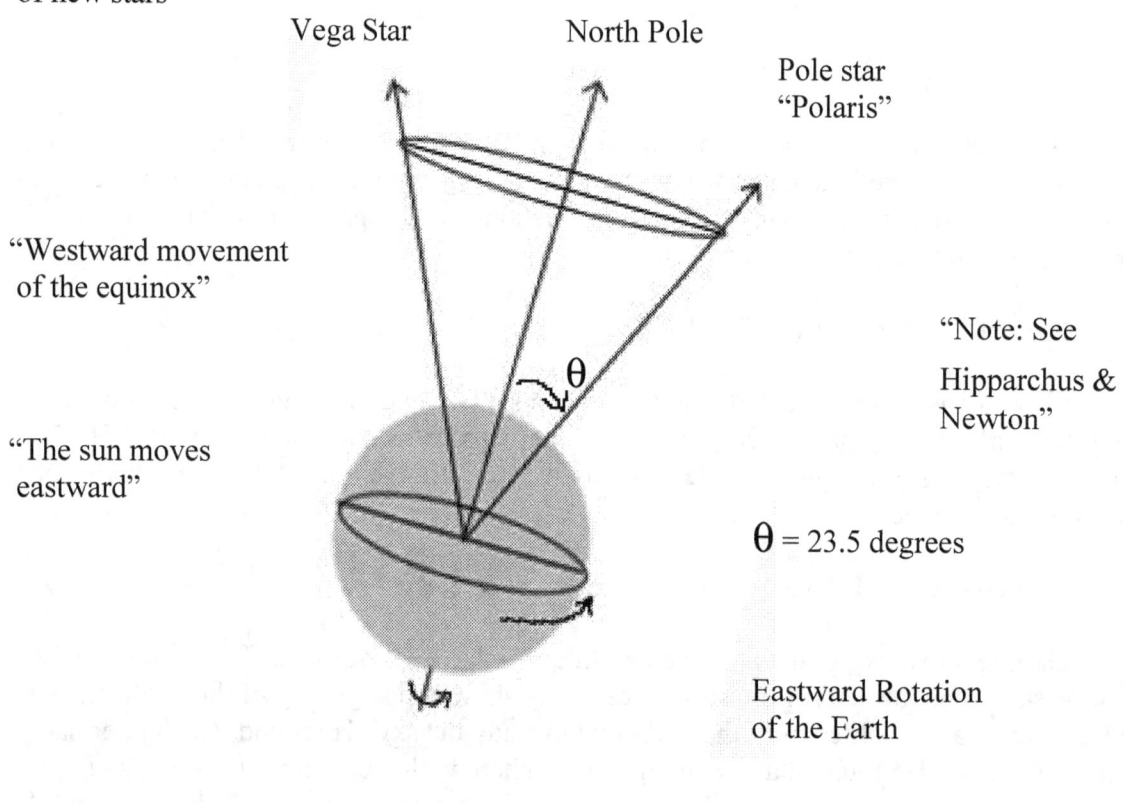

"Precessional motion
of new stars"

Vega Star   North Pole

Pole star
"Polaris"

"Westward movement
of the equinox"

"Note: See

Hipparchus &
Newton"

θ

"The sun moves
eastward"

θ = 23.5 degrees

Eastward Rotation
of the Earth

Earth

Spring Equinox: March 21
Autumn Equinox: September 23

Notes

1)   The star's movement is westward as the earth's rotation is eastward. The motion of

the new pole star's westward movement on the Ecliptic is due to the precession of the

equinoxes.

2)   It takes 26,000 years for the precession to complete. In 13,000 years the Vega Star will become the new Pole Star.

3)   Hipparchus (527 B.C. to 514 B.C.) was the first astronomer to note the celestial rotation of the new pole stars from the records of past ages. He found a westward rotation of the stars. Our present pole star is Polaris. The celestial equator moves along the ecliptic westerly at the rate of 50 arc seconds per 26,000 years. Sir Isaac Newton calculated the annual westward motion to be about 50 seconds per arc.

How the Area Constant was Developed

$$E = [\hbar(C/\lambda)] = [(\mathcal{A}_p C^2/G)C^2]$$

Set: $\left[\dfrac{(\hbar C)\ G}{C^4}\right] = [(\mathcal{A}_p \times \lambda_p)]$

Set: $[(\mathcal{A}_p \times \lambda_p)] = [(\hbar/C^3)\ G]$ <u>Final Concept</u>

Data:

$\hbar = 6.6260755 \times 10^{-34}$ joules*s, Max Planck's Constant

$C = 2.99792458 \times 10^8$ m/s, velocity of light

$C^3 = 2.69440024174 \times 10^{25}$ m^3/s^3

$\mathbf{P_c} = (\hbar/C^3) = 2.45920238477 \times 10^{-59}$ kg s^2/m, Pershing's Constant

Note:

This universal area constant may always be used for any planet or any given mass of any object. Method of how to find the gravitational structural length is through Einstein's work and the bending of space.

I set my Energy equation equal to Max Planck's energy equation. My equation of energy is [$(\mathcal{A}_p C^2/G)C$]. Max Planck's is [$\hbar(C/\lambda)$].

Pershing's Universal Area Constant

$[\,(\hbar/\,C^3\,)\,G\,] = 1.64092492406 \times 10^{-69}\ m^2$, "The Universal Area Constant"

Data:

$\hbar = 6.6260755 \times 10^{-34}\ kg\ m^2/s$, Max Planck's Constant

$C = 2.99792458 \times 10^8\ m/s$, velocity of light

$C^3 = 2.69440024174 \times 10^{25}\ m^3/s^3$

$G = 6.67259 \times 10^{-11}\ m^3/\,kg\ s^2$, Isaac Newton's Gravitational Constant

$\mathbf{P}_c = (\hbar/\,C^3\,) = 2.45920238477 \times 10^{-59}\ kg\ s^2/m$

A New Concept of Gravitational Structural Length
to determine one AU

$$a_s = \left[(\mathcal{A}_E + \mathcal{A}_s) \left[\frac{C \, P_E}{2\pi} \right]^2 \right]^{1/3}$$

$a_s = 1.4959787 \times 10^{11}$ m = 1 AU Exact

Data:

Gravitational Structural Length of the Sun and Earth
$(\mathcal{A}_E + \mathcal{A}_s) = 1{,}476.62936702$ m
$C = 2.99792458 \times 10^8$ m/s, velocity of light
(Tropical Year) x (Mean Solar Year)
$P_s = 365.256360417$ days x $\dfrac{86400"}{1 \text{ day}}$

$P_s = 31558149.54$ s

Mean Solar Year:

365 days x $\dfrac{86400"}{1 \text{ day}}$	=	31536000.00"
6 hr x $\dfrac{3600"}{1 \text{ hr}}$	=	21600.00"
9 min x $\dfrac{60"}{1 \text{ min}}$	=	540.00"
9.54 sec	=	9.54"
* Total	=	31558149.54 sec

$$\left[\frac{C \, P_E}{2\pi} \right] = 1.50574824042 \times 10^{15} \text{ m} \qquad 2\pi = 6.2831853071$$

$$\left[\frac{C \, P_E}{2\pi} \right]^2 = 2.26727776353 \times 10^{30} \text{ m}$$

* AU: Astronomical Unit – The mean distance of the Earth from the Sun.

Gravitational Structural Length Concepts

1) $\mathcal{A}_{\text{deuteron}} = 5.63308491407 \times 10^{-15}$ m

2) $\mathcal{A}_{\text{neutron}} = 5.63308491407 \times 10^{-15}$ m

3) $\mathcal{A}_{\text{muon}} = 3.17326954878 \times 10^{-16}$ m

4) $\mathcal{A}_{\text{proton}} = 2.81794096266 \times 10^{-15}$ m

5) $\mathcal{A}_{\text{electron}} = 1.53469854508 \times 10^{-18}$ m

----

1) $\mathcal{A}_{\text{deutron}} \, C^2 = 506.276423879$ m^3/s^2

2) $\mathcal{A}_{\text{neutron}} \, C^2 = 253.612995706$ m^3/s^2

3) $\mathcal{A}_{\text{muon}} \, C^2 = 28.5199244049$ m^3/s^2

4) $\mathcal{A}_{\text{proton}} \, C^2 = 253.263903357$ m^3/s^2

5) $\mathcal{A}_{\text{electron}} \, C^2 = 0.137931826519$ m^3/s^2

Where $C^2 = 8.98755178737 \times 10^{16}$ m^2/s^2

$M_{\text{electron}} \, (G_P) = 0.137931826519$ m^3/s^2

$G_P = 1.51417198146 \times 10^{29}$ m^3/ kg s^2,

        Pershing's Gravitational Constant of the Hydrogen Atom

Mass of electron $= 9.10938970 \times 10^{-31}$ kg

Concepts of $\mathrm{M}\, G = \mathcal{A}\, C^{2}$

Earth Concepts

$M_{\mathbf{E}} = 5.97506244160 \times 10^{24}$ kg
$G = 6.67259 \times 10^{-11}$ m^3/ kg s^2

$M_{\mathbf{E}}\, G = 3.98691418972 \times 10^{14}$ m^3/s^2

$\mathcal{A}_{\mathbf{E}} = 4.43604029667 \times 10^{-3}$ m, Gravitational Structural Length of the Earth
$C^{2} = 8.98755178737 \times 10^{16}$ m^2/s^2

$\mathcal{A}_{\mathbf{E}}\, C^{2} = 3.98691418972 \times 10^{14}$ m^3/s^2

Sun Concepts

$M_{\mathbf{S}} = 1.9889193184 \times 10^{30}$ kg
$G = 6.67259 \times 10^{-11}$ m^3/ kg s^2

$M_{\mathbf{S}}\, G = 1.32712431548 \times 10^{20}$ m^3/s^2

$\mathcal{A}_{\mathbf{S}} = 1476.62494401$ m, Gravitational Structural Length of the Earth
$C^{2} = 8.98755178737 \times 10^{16}$ m^2/s^2

$\mathcal{A}_{\mathbf{S}}\, C^{2} = 1.32712431548 \times 10^{20}$ m^3/s^2

Introduction of the Hydrogen Atom

M_p = Mass of the Proton = $1.6726231 \times 10^{-27}$ kg
M_e = Mass of the Electron = $9.1093897 \times 10^{-31}$ kg

Ratio of $\dfrac{M_p}{M_e} = \dfrac{1.6726231 \times 10^{-27} \text{ kg}}{9.1093897 \times 10^{-31} \text{ kg}} = 1,836.15275566$

$\mathcal{A}_p$ = $2.81794096266 \times 10^{-15}$ m, Gravitational Structural Length of the Proton
$\mathcal{A}_e$ = $1.53469854508 \times 10^{-18}$ m, Gravitational Structural Length of the Electron

Ratio of $\dfrac{\mathcal{A}_p}{\mathcal{A}_e} = \dfrac{2.81794096266 \times 10^{-15} \text{ m}}{1.53469854508 \times 10^{-18} \text{ m}} = 1,836.15275566$

Therefore, $\dfrac{M_p}{M_e} = \dfrac{\mathcal{A}_p}{\mathcal{A}_e}$

Pershing's Area Constant of the Hydrogen Atom

Area Constant of Hydrogen Atom $= (\hbar / C^3) \, G_P = (\mathcal{A}_p \times \lambda_p) = 3.72365534776 \times 10^{-30} \text{ m}^2$

Data:

$\hbar = 6.6260755 \times 10^{-34}$ joules s

$C = 2.99792458 \times 10^8$ m/s

$C^3 = 2.69440024174 \times 10^{25}$ m^3/s^3

$P_{c_1} = (\hbar / C^3) = 2.45920238477 \times 10^{-59}$ kg s^2/m, Pershing's First Constant

$G_P = 1.51417198146 \times 10^{29}$ m^3/ kg s^2,
      Pershing's Gravitational Constant of the Hydrogen Atom

$\mathcal{A}_p = 2.81794096266 \times 10^{-15}$ m, Gravitational Structural Length of the proton

$(\hbar / C^3) \, G_P = 3.72365534776 \times 10^{-30}$ m^2

$\lambda_p = [(\hbar / C^3) \, G_P] / \mathcal{A}_p = 1.32140999301 \times 10^{-15}$ m, wavelength of the proton

Therefore:

$f_p = (C / \lambda_p) = 2.26873157904 \times 10^{23}$ vib/s

$E_p = \hbar f_p = 1.5032786732 \times 10^{-10}$ joules

Pershing's Ratio Concepts

Data:

$\mathcal{A}_p C^2 = M_p G_P$
$\mathcal{A}_p / M_p = G_P / C^2$

$\dfrac{\mathcal{A}_p}{M_p} \dfrac{= 2.81794096266 \times 10^{-15} \text{ m}}{1.6726231 \times 10^{-27} \text{ kg}}$

$\dfrac{\mathcal{A}_p}{M_p} = 1.68474354005 \times 10^{12}$ m/kg

$\dfrac{G_P}{C^2} \dfrac{= 1.51417198146 \times 10^{29} \text{ m}^3/ \text{ kg s}^2}{8.98755178737 \times 10^{16} \text{ m}^2/\text{s}^2}$

$\dfrac{G_P}{C^2} = 1.68474354005 \times 10^{12}$ m/kg

Therefore, $\mathcal{A}_p / M_p = G_P / C^2$

$E_p = M_p\, C^2 = \hbar f_p$, where $C^2 = 8.98755178737 \times 10^{16}$ m²/s²

Therefore, $M_p = (\hbar f_p)/ C^2$

$M_p = \dfrac{1.5032786732 \times 10^{-10} \text{ joules}}{8.89755178737 \times 10^{16} \text{ m}^2/\text{s}^2}$

$M_p = 1.6726231 \times 10^{-27}$ kg

Pershing's Gravitational Constant of Hydrogen Atom

- First Concept

$$G_P = \frac{A_p C^2}{M_p} = 1.51417198146 \times 10^{29} \text{ m}^3/\text{ kg s}^2$$

Data:

$A_p = 2.81794096266 \times 10^{-15}$ m, Gravitational Structural Length of the proton

$C^2 = 8.98755178737 \times 10^{16} \text{ m}^2/\text{s}^2$

$A_p C^2 = 253.263903357 \text{ m}^3/\text{s}^2$

$M_p = 1.6726231 \times 10^{-27}$ kg, mass of proton

- Second Concept

$$G_P = \frac{A_e C^2}{M_e} = 1.51417198146 \times 10^{29} \text{ m}^3/\text{ kg s}^2$$

Data:

$A_e = 1.53469854508 \times 10^{-18}$ m, Gravitational Structural Length of the electron

$C^2 = 8.98755178737 \times 10^{16} \text{ m}^2/\text{s}^2$

$A_e C^2 = 0.137931826519 \text{ m}^3/\text{s}^2$

$M_e = 9.10938970 \times 10^{-31}$ kg, mass of proton

- Third Concept

$$G_P = N_p\, G = 1.51417198146 \times 10^{29} \text{ m}^3/\text{ kg s}^2$$

Data:

$N_p = 2.2694175089 \times 10^{39}$, Pershing's constant of the Proton
$G = 6.67259 \times 10^{-11} \text{ m}^3/\text{kg s}^2$, Newton's Gravitational Constant

$G_P = 1.51417198146 \times 10^{29} \text{ m}^3/\text{ kg s}^2$,

Pershing's Gravitational Constant of the Hydrogen Atom

- Fourth Concept:

$(M_p \, N_p) \, G = 253.263903357 \text{ m}^3/\text{s}^2$

Data:

$M_p = 1.6726231 \times 10^{-27}$ kg, mass of the proton

$N_p = 2.2692417509 \times 10^{12}$, Pershing's constant of the Proton

$(M_p \, N_p) = 3.79558617204 \times 10^{12}$ kg

$G = 6.67259 \times 10^{-11} \text{ m}^3/\text{kg s}^2$, Newton's Gravitational Constant

- Fifth Concept:

$(E_p / \, C^2) \, (N_p) \, G = 253.263903357 \text{ m}^3/\text{s}^2$

Data:

$E_p = M_p \, C^2 = 1.5032786732 \times 10^{-10} \text{ kg m}^2/\text{s}^2$

$M_p = 1.6726231 \times 10^{-27}$ kg, mass of the proton

$C^2 = 8.98755178737 \times 10^{16} \text{ m}^2/\text{s}^2$

$(E_p / \, C^2) = M_p = 1.6726231 \times 10^{-27}$ kg

$(E_p / \, C^2) \, (N_p) = 3.79558617204 \times 10^{12}$ kg

$N_p = 2.2692417509 \times 10^{12}$, Pershing's constant of the Proton

$(E_p / \, C^2) \, (N_p) \, G = 253.263903357 \text{ m}^3/\text{s}^2$

- Sixth Concept:

$\mathcal{A}_e \, (M_p/M_e) \, C^2 = 253.263903357 \; m^3/s^2$

$\dfrac{\mathcal{A}_p}{\mathcal{A}_e} = \dfrac{(2.81794096266 \times 10^{-15} \; m)}{(1.53469854508 \times 10^{-18} \; m)}$

$\dfrac{\mathcal{A}_p}{\mathcal{A}_e} = 1{,}836.15275566$

$\dfrac{M_p}{M_e} = \dfrac{1.6726231 \times 10^{-27} \; kg}{9.10938970 \times 10^{-31} \; kg}$

$\dfrac{M_p}{M_e} = 1{,}836.15275566$

$C^2 = 8.98755178737 \times 10^{16} \; m^2/s^2$

$\mathcal{A}_e \, (M_p/M_e) \, C^2 = 253.263903357 \; m^3/s^2$

16

Ratio Concepts of the Hydrogen Atom

A) $\dfrac{\text{mass of the deuteron}}{\text{mass of the electron}} = \dfrac{3.3435860 \times 10^{-27} \text{ kg}}{9.1093897 \times 10^{-31} \text{ kg}} = 3{,}670.48299624$

Gravitational Structural Length Ratios of the deutron and electron

B) $\dfrac{\mathcal{A}\text{ deuteron}}{\mathcal{A}\text{ electron}} = \dfrac{5.63308491407 \times 10^{-15} \text{ m}}{1.53469854508 \times 10^{-18} \text{ m}} = 3{,}670.48299624$

C) $\dfrac{\text{mass of the neutron}}{\text{mass of the electron}} = \dfrac{1.6749286 \times 10^{-27} \text{ kg}}{9.1093897 \times 10^{-31} \text{ kg}} = 1{,}838.68366066$

Gravitational Structural Length Ratios of the neutron and electron

D) $\dfrac{\mathcal{A}\text{ neutron}}{\mathcal{A}\text{ electron}} = \dfrac{2.82182513889 \times 10^{-15} \text{ m}}{1.53469854508 \times 10^{-18} \text{ m}} = 1{,}838.68366066$

E) $\dfrac{\text{mass of muon}}{\text{mass of electron}} = \dfrac{1.88353270 \times 10^{-28} \text{ kg}}{9.10938970 \times 10^{-31} \text{ kg}} = 206.768264618$

Gravitational Structural Length Ratios of the muon and electron

F) $\dfrac{\mathcal{A}\text{ muon}}{\mathcal{A}\text{ electron}} = \dfrac{3.17326954878 \times 10^{-16} \text{ m}}{1.53469854508 \times 10^{-18} \text{ m}} = 206.768264618$

G) $\dfrac{\text{mass of proton}}{\text{mass of electron}} = \dfrac{1.6726231 \times 10^{-27} \text{ kg}}{9.10938970 \times 10^{-31} \text{ kg}} = 1{,}836.15275566$

Gravitational Structural Length Ratios of the proton and electron

H) $\dfrac{\mathcal{A}\text{ proton}}{\mathcal{A}\text{ electron}} = \dfrac{2.81794096266 \times 10^{-15} \text{ m}}{1.53469854508 \times 10^{-18} \text{ m}} = 1{,}836.15275566$

Note: The gravitational structural lengths give a more exact value.

Energy Concept of the Hydrogen Atom

$$E = \frac{-(\hbar R_\infty)}{n^2} = \frac{(P_c)(R_\infty)}{n^2}$$

$E = -2.17987413195 \times 10^{-18}$ joules, for the first shell of the hydrogen atom
$E = -5.44968532988 \times 10^{-19}$ joules, for the second shell of the hydrogen atom

If n=1, it is the energy of the first shell.
If n=2, it is the energy of the second shell.

$P_c = (\hbar)(C) = 1.98644746104 \times 10^{-25}$ joules*meter

$\hbar = 6.6260755 \times 10^{-34}$ joules s, Max Planck's constant
$C = 2.99792458 \times 10^{8}$ m/s, velocity of light
$R_\infty = 10,973,731.6224$ /meter, Rydberg constant

$$\frac{(E_{1st\ shell})}{(E_{2nd\ shell})} = \frac{-2.17987413195 \times 10^{-18}\ joules}{-5.44968532988 \times 10^{-19}\ joules}$$

$$\frac{(E_{1st\ shell})}{(E_{2nd\ shell})} = 4 = 2^2$$

$KE_e = \frac{1}{2}(M_e)(v_e)^2$
$KE_e = 2.17987413195 \times 10^{-18}$ joules

$(v_e) = 2,187,691.43195$ m/s, velocity of electron in 1st shell of Hydrogen atom
$(v_e)^2 = 4.78599380143 \times 10^{12}$ m^2/s^2
$(M_e) = 9.1093897 \times 10^{-31}$ kg, mass of electron

$$E_{1st\ shell} = -2.17987413195 \times 10^{-18}\ joules \ \times \ \frac{1\ eV}{1.60217733 \times 10^{-19}\ joules}$$

$E_{1st\ shell} = -13.6056982653$ eV , where n =1 in the 1st energy shell

$$E_{2nd\ shell} = -5.44968532988 \times 10^{-19}\ joules \ \times \ \frac{1\ eV}{1.60217733 \times 10^{-19}\ joules}$$

$E_{2nd\ shell} = -3.40142456632$ eV , where n =2 in the 2nd energy shell

Therefore, $\dfrac{-13.6056982653\ eV}{-3.40142456632\ eV} = 4 = 2^2$

Centripetal Force & Centrifugal Force of the Hydrogen Atom

Centripetal Force - "inward force"

$$F_{inward} = \left(\frac{A_p \; C^2 \; M_e}{(a_o)^2} \right)$$

$F_{inward} = 8.23872959795 \times 10^{-8}$ kg m/s^2

Data:

$A_p = 2.81794096266 \times 10^{-15}$ m, Gravitational Structural Length of Proton

$C = 2.99792458 \times 10^8$ m/s, velocity of light

$C^2 = 8.98755178737 \times 10^{16}$ m^2/s^2

$A_p \; C^2 = 253.263903357$ m^3/s^2

$M_e = 9.1093897 \times 10^{-31}$ kg, mass of electron

$a_o = 5.29177249 \times 10^{-11}$ m

$(a_o)^2 = 2.80028560859 \times 10^{-21}$ m^2

$[(A_p \; C^2) \; M_e] = 2.30707959262 \times 10^{-28}$ kg m^3/s^2

$[(A_p \; C^2) \; M_e] / (a_o)^2 = 8.23872959795 \times 10^{-8}$ kg m/s^2

Centrifugal Force – "outward force"

$$F_{outward} = \left(\frac{M_e \, v_e^2}{a_o} \right)$$

$F_{outward} = 8.23872959795 \times 10^{-8}$ kg m/s^2

Data:

$M_e = 9.10938970 \times 10^{-31}$ kg, mass of electron

$v_e = 2187691.43195$ m/s, average orbital velocity of electron

$v_e^2 = 4.78599380143 \times 10^{12}$ m^2/s^2

$M_e \, v_e^2 = 4.3597482639 \times 10^{-18}$ kg m^2/s^2

$a_o = 5.29177249 \times 10^{-11}$ m, Bohr's radius

The Gravitational Structural Length & Three Energy Concepts of Water

A) The Mass of Water

$$H = 1.00794 \text{ amu} \times \frac{1.6605402 \times 10^{-27} \text{ kg}}{1 \text{ amu}}$$

$$H = 1.67372488919 \times 10^{-27} \text{ kg}$$

$$H^2 = 2.80135500469 \times 10^{-54} \text{ kg}$$

* amu = atomic mass unit

Oxygen Mass:

$$O = 15.9994 \text{ amu} \times \frac{1.6605402 \times 10^{-27} \text{ kg}}{1 \text{ amu}}$$

$$O = 2.65676468759 \times 10^{-26} \text{ kg}$$

Mass of Water:

$$M_{water} = (H_2O) = 2.65676468759 \times 10^{-26} \text{ kg}$$

B) <u>First Energy Concept of Water</u>: Albert Einstein Concept

$$E_w = M_w \, C^2 \quad \text{where } w = H_2O$$

$$E_w = 2.38778102166 \times 10^{-9} \text{ joules}$$

Where $C = 2.99792458 \times 10^8$ m/s & $C^2 = 8.98755178737 \times 10^{16}$ m^2/s^2

C) $M_{water} \, G = 1.77275014868 \times 10^{-36}$ m^3/s^2

Data:

$$M_{water} = 2.65676468759 \times 10^{-26} \text{ kg}$$

$G = 6.67259 \times 10^{-11}$ m^3/kg s^2 , Sir Isaac Newton's Gravitational Constant

D) $\mathcal{A}_{water} \, C^2 = 1.77275014868 \times 10^{-36}$ m^3/s^2

Data:

$\mathcal{A}_{water} = 1.9724505523 \times 10^{-53}$ m, Gravitational Structural Length of Water

$C = 2.99792458 \times 10^8$ m/s

$C^2 = 8.98755178737 \times 10^{16}$ m^2/s^2

Therefore, $M_{water} \, G = \mathcal{A}_{water} \, C^2$

E) <u>Second Energy Concept of Water</u>: D. Pershing's Concept

$$E_w = \frac{\mathcal{A}_w\,C^4}{G}$$

$$E_w = 2.38778102167 \times 10^{-9}\ \text{joules}$$

Data:

$C = 2.99792458 \times 10^8\ \text{m/s}$

$C^4 = 8.07760871307 \times 10^{33}\ \text{m}^4/\text{s}^4$

$G = 6.67259 \times 10^{-11}\ \text{m}^3/\text{kg s}^2$

$C^4/G = 1.21056571932 \times 10^{44}\ \text{kg m/s}^2$

$\mathcal{A}_w = 1.9724505523 \times 10^{-53}\ \text{m}$, Gravitational Structural Length of Water

$E_w = \dfrac{\mathcal{A}_w\,C^4}{G} = 2.38778102167 \times 10^{-9}\ \text{joules}$

F) <u>Third Energy Concept of Water</u>: Max Planck's Concept

$$E_w = \frac{\hbar\,C}{\lambda_w}$$

$$E_w = \hbar\,\nu_{water} = 2.38778102166 \times 10^{-9}\ \text{joules}$$

Data:

$\hbar = 6.6260755 \times 10^{-34}\ \text{joule*s}$

$C = 2.99792458 \times 10^8\ \text{m/s}$

$\lambda_w = 8.31921957256 \times 10^{-17}\ \text{m}$

$\nu_{water} = \dfrac{C}{\lambda_w} = 3.60361275941 \times 10^{24}\ \text{vib/s}$

Area Constant of Water

A) Pershing's Area Constant

$$\mathcal{A}_w \times \lambda_w = 1.64092492406 \times 10^{-69} \text{ m}^2$$

Data:

$$\mathcal{A}_w = 1.9724505523 \times 10^{-53} \text{ m}$$

$$\lambda_w = 8.31921957256 \times 10^{-17} \text{ m}$$

B) Pershing's Second Area Constant

$$(\hbar / C^3) \, G = 1.64092492406 \times 10^{-69} \text{ m}^2$$

Data:

$\hbar = 6.6260755 \times 10^{-34}$ joule*s, Planck's constant

$C = 2.99792458 \times 10^8$ m/s

$C^3 = 2.69440024174 \times 10^{25}$ m^3/s^3

$(\hbar / C^3) = 2.45920238477 \times 10^{-59}$ kg s^2/m

$G = 6.67259 \times 10^{-11}$ m^3/kg s^2, Sir Isaac Newton's Gravitational Constant

Introduction of the Nature of the Universe

Three Energy Concepts

1) $E = MC^2$ Einstein's Energy Concept
2) $E = (\hbar C)/\lambda$ Max Planck's Energy Concept
3) $E = (A C^4)/G$ D. Pershing's Energy Concept

Data:

$\hbar C = 1.98644746104 \times 10^{-25}$ joules m
$\hbar = 6.6260755 \times 10^{-34}$ joules s, Max Planck's Constant
$C = 2.99792458 \times 10^8$ m/s, velocity of light

$C^4 = 8.07760871306 \times 10^{33}$ m^4/s^4
$G = 6.67259 \times 10^{-11}$ m^3/kg s^2, Isaac Newton's Gravitational Constant
$\mathbf{P}_{c_2} = (C^4/G) = 1.21056571932 \times 10^{44}$ kg m/s^2, Pershing's 2nd Concept

Pershing's Universal Area Constant
(based on a Universal Momentum Constant of Nature)

$MC = (\hbar)/\lambda = (A C^3)/G$
I believe this is what Albert Einstein was looking for in the nature of the universe.

Therefore,

$A \times \lambda = (\hbar/C^3) \times G$
$A \times \lambda = 1.64092492406 \times 10^{-69}$ m^2

Data:

$C^3 = 2.69440024174 \times 10^{25}$ m^3/s^3
$\mathbf{P}_{c_1} = (\hbar/C^3) = 2.45420238477 \times 10^{-59}$ kg s^2/m

Three Energy Concepts Based on Momentum

1) MC = momentum
$MC = (\hbar)/\lambda = (A C^3)/G$, three concepts of momentum - constant always

2) Energy Universal Concepts
$E = MC^2 = (\hbar C)/\lambda = (A C^4)/G$

3) $MG = A C^2$ this is always true in nature

4) if $MG = \mathcal{A} C^2$, then $M = (\mathcal{A} C^2)/G$
   * $E = MC^2 = (\mathcal{A} C^4)/ G$: Basic Concept of Nature

5) if $MC = (\hbar)/ \lambda$,
   then $E = MC^2 = (\hbar C)/ \lambda$: Basic Concept of Nature

6) Therefore, $E = MC^2 = (\hbar C)/ \lambda = (\mathcal{A} C^4)/ G$

D. Pershing's Ratio Constant

- First Ratio Constant

$\mathcal{A} / M = G/ C^2 = 7.42425763752 \times 10^{-28}$ m/kg

Data:

$G = 6.67259 \times 10^{-11}$ m^3/kg s^2, Isaac Newton's Gravitational Constant
$C = 2.99792458 \times 10^8$ m/s, velocity of light
$C^2 = 8.98755178737 \times 10^{16}$ m^2/s^2
$G/ C^2 = 7.42425763752 \times 10^{-28}$ m/kg
$\mathcal{A}$ = Gravitational Structural Length, the distortion of space caused by the mass
G = Newton's Gravitational Structural Length

Note:

7.42425763752 $\times 10^{-28}$ m/kg using (1/x) on calculator, we
obtain: 1.34693601546 $\times 10^{27}$ kg/m

- Second Ratio Constant

$M/\mathcal{A} = C^2/G = 1.34693601546 \times 10^{27}$ kg/m

Data:

$G = 6.67259 \times 10^{-11}$ m^3/kg s^2, Isaac Newton's Gravitational Constant
$C = 2.99792458 \times 10^8$ m/s, velocity of light
$C^2 = 8.98755178737 \times 10^{16}$ m^2/s^2

Note:

$(M/\mathcal{A}) \times (\mathcal{A}/ M) = 1$

Energy Concepts of the Earth

First Energy Concept - Einstein's Energy concept

$$E_E = M_E C^2$$
$$E_E = 5.35433958929 \times 10^{41} \text{ Joules}$$

Data:

$$M_E = 5.957506244160 \times 10^{24} \text{ kg}$$
$$C^2 = 8.98755178737 \times 10^{16} \text{ m}^2/\text{s}^2$$

Second Energy Concept - D. Pershing's Energy Concept

$$E_E = \frac{A_E C^4}{G} = 5.35433958929 \times 10^{41} \text{ joules,}$$

Data:

$$A_E = 4.42300612337 \times 10^{-3} \text{ m, Gravitational Structural Length of the Earth}$$
$$C^4 = 8.07760871307 \times 10^{33} \text{ m}^4/\text{s}^4$$
$$G = 6.67259 \times 10^{-11} \text{ m}^3/ \text{ kg s}^2 \text{, Sir Isaac Newton's Gravitational Constant}$$

$$P_c = \frac{C^4}{G} = 1.21056571932 \times 10^{44} \text{ kg m/s}^2 \text{, D. Pershing's Energy Concept}$$

Third Energy Concept – Max Planck's Energy Concept

$$E_E = \hbar C/\lambda_E = \hbar \nu_E$$
$$E_E = 5.35433958929 \times 10^{41} \text{ Joules}$$

Data:
$$C = 2.99792458 \times 10^8 \text{ m/s, velocity of light}$$
$$\lambda_E = 3.69935426786 \times 10^{-67} \text{ m, wavelength of earth}$$
$$\nu_E = C/\lambda_E = 8.10452327726 \times 10^{74} \text{ vibrations/s}$$
$$\hbar = 6.6260755 \times 10^{-34} \text{ Joules} * \text{s, max Planck's constant}$$
$$E_E = \hbar \nu_E = 5.37011831266 \times 10^{41} \text{ joules}$$

Angular Rotation of the Earth
& Sun and Earth System

- First Concept

  P_{earth} = 31558149.54 s, one year period of Earth

  $\omega = (2\pi/ P) = 1.99098660687 \times 10^{-7}$ rad/s

  <u>Check:</u>
  $\omega P = 2\pi = 6.28318530718$

- Second Concept

  $\omega = (v_{earth} / a_{sun}) = 1.99098660687 \times 10^{-7}$ rad/s

  v_{earth} = 29,784.7355586 m/s
  $(v_{earth})^2$ = 887,130,472.296 m^2/s^2
  $a_{sun} = 1.4959787 \times 10^{11}$ m

- Third Concept

  $\omega^2 = 3.96402766874 \times 10^{-14}$ rad/s^2
  $a^3 = 3.34792892881 \times 10^{33}$ m^3

  $a^3 \omega^2 = 1.32712829068 \times 10^{20}$ m^3/s^2
  $v^2 a = 1.32712829068 \times 10^{20}$ m^3/s^2

- Fourth Concept

  $M_{sun} = 1.9889193184 \times 10^{30}$ kg
  $M_{earth} = 5.957506244160 \times 10^{24}$ kg

  $M_{sun} + M_{earth} = 1.98892527591 \times 10^{30}$ kg

  <u>Check:</u>
  $G = 6.67259 \times 10^{-11}$ m^3/kgs^2 , Newton's gravitational constant
  $[(M_{sun} + M_{earth}) \, G] = 1.32712829068 \times 10^{20}$ m^3/s^2

- Fifth Concept

$$(4\pi^2)\,(a^3/P^2) = 1.32712829068 \times 10^{20}\ \text{m}^3/\text{s}^2$$

Data:
$\pi = 3.14159265359$
$\pi^2 = 9.86960440109$
$4\pi^2 = 39.4784176044$
$a_{sun} = 1.4959787 \times 10^{11}\ \text{m}$
$a^3 = 3.34792892881 \times 10^{33}\ \text{m}^3$
$P_{earth} = 365.256360417\ \text{days} \times (86400\ \text{s}/\ 1\ \text{day})$
$P_{earth} = 31558149.54\ \text{s}$
$P^2 = 9.95916802389 \times 10^{14}\ \text{s}^2$
$(a^3/P^2) = 3.36165523142 \times 10^{18}\ \text{m}^3/\text{s}^2$

$$(4\pi^2)\,(a^3/P^2) = 1.32712829068 \times 10^{20}\ \text{m}^3/\text{s}^2$$

- Final Concept

$$(4\pi^2)\,(a^3/P^2) = [(\mathcal{A}_{sun} + \mathcal{A}_{earth})\,C^2] = [(M_{sun} + M_{earth})\,G] = v^2\,a = a^3\,\omega^2$$

Perturbation of Earth

$$\Delta \omega_{Earth} = \left(\frac{3}{(1-e^2)}\right) \left(\frac{v_{Earth}}{C}\right)^2 \left(\frac{1,296,000"}{1 \text{ yr}}\right) \left(\frac{100 \text{ yr}}{\text{century}}\right)$$

$\Delta \omega_{Earth} = 3.83878231072$ seconds of arc/century

Data:

$v_{Earth} = 29,784.7355587$ m/s, orbital velocity of Earth

$C = 2.99792458 \times 10^8$ m/s, velocity of light

$(v_{Earth} / C) = 9.93511836735 \times 10^{-5}$ rad

$(v_{Earth} / C)^2 = 9.87065769733 \times 10^{-9}$ rad

$e_{Earth} = 0.0167$, eccentricity of the Earth

$(1-e^2) = 0.99972111$

Two Basic Concepts of Centripetal Force of the Earth

1)   First Concept of Centripetal Force

$$F_{Earth} = \frac{M_{Earth}\,(V_{Earth})^2}{a_{Earth}}$$ Average centripetal force

$F_{Earth} = 3.53286134896 \times 10^{22}$ kg m/s$^2 \approx$ nt

Data:

$M_{Earth} = 5.95750624416 \times 10^{24}$ kg

$V_{Earth} = 29{,}784.7355586$ m/s

$(V_{Earth})^2 = 887{,}130{,}472.296$ m^2/s^2

$a_{Earth} = 1.4959787 \times 10^{11}$ m, distance to the sun

2)   Second Concept of Centripetal Force

$$F_{Earth} = \frac{M_{Earth}\,[(\mathcal{A}_{Earth} + \mathcal{A}_{moon})C^2]}{(a_{Earth})^2}$$ Average centripetal force

$F_{Earth} = 3.53286134896 \times 10^{22}$ kg m/s$^2 \approx$ nt

Data:

$(a_{Earth})^2 = 2.23795227085 \times 10^{22}$ m^2

$M_{Earth} = 5.95750624416 \times 10^{24}$ kg

$(\mathcal{A}_{Earth} + \mathcal{A}_{moon}) = 1{,}476.62936702$ m

$C^2 = 8.98755178737 \times 10^{16}$ m^2/s^2

$(\mathcal{A}_{Earth} + \mathcal{A}_{moon})C^2 = 1.32712829068 \times 10^{20}$ m^3/s^2

Escape Velocity of the Earth

$$v_{escape} = C \left[\frac{(2\,\mathcal{A}_{Earth})}{(\mathcal{R}_{Earth})\,(1 - 0.0167^2)} \right]^{\frac{1}{2}} = 2,372.04075713 \text{ m/s}$$

$$v_{escape} = 11,166.2671965 \text{ m/s} \times \frac{1 \text{ km}}{1,000 \text{ m}}$$

$$v_{escape} = 11.166267195 \text{ km/s}$$

$\mathcal{A}_{Earth} = 4.42300612337 \times 10^{-3} \text{ m}$
$\mathcal{R}_{Earth} = 6,378,140 \text{ m}$

Ellipse of the Earth

First Method for Orbital Velocity of Earth

$$v_{Earth} = \frac{2\pi a_{Earth}}{P_{Earth}}$$

$$v_{Earth} = \frac{2\pi(1.4959787 \times 10^{11} m)}{31558149.540\ s}$$

$v_{Earth} = 29,784.7355586$ m/s, average orbital velocity of Earth

$v_{perigee} = v(1+e)$, where $e_{Earth} = 0.0167$
$v_{perigee} = 30,282.1406424$ m/s

$v_{apogee} = v(1-e)$
$v_{apogee} = 29,287.330478$ m/s

$$v_{Earth} = \frac{(v_{perigee} + v_{apogee})}{2}$$

$$v_{Earth} = \frac{59,569.4711172\ m/s}{2}$$

$v_{Earth} = 29,784.7355586$

Second Method for Orbital Velocity of Earth

$$v_{Earth} = C\sqrt{\frac{(\mathcal{A}_{Earth} + \mathcal{A}_{sun})}{a_{Earth}}}$$

$$\frac{(\mathcal{A}_{Earth} + \mathcal{A}_{sun})}{a_{Earth}} = \frac{1476.62936702\ m}{1.4959787 \times 10^{11}\ m} = 9.87065769733 \times 10^{-9}\ rad$$

$$\sqrt{9.87065769733 \times 10^{-9}\ rad} = 29,784.7355586\ m/s$$

$v_{Earth} = 29,784.7355586$ m/s

Three Energy Concepts of the Moon

1)   Albert Einstein's Energy Concept

$E_{Moon} = [(M_{Moon}) C^2]$
$E_{Moon} = 6.58585317651 \times 10^{39}$ joules

$M_{Moon} = 7.327749906 \times 10^{22}$ kg
$C^2 = 8.98755178737 \times 10^{16}$ m^2/s^2

2)   Max Planck's Energy Concept

$E_{Moon} = (\hbar)(\nu_{Moon}) = \hbar (C / \lambda_{Moon})$
$E_{Moon} = 6.58583788293 \times 10^{39}$ joules

$\hbar = 6.6260755 \times 10^{-34}$ Joules * s, max Planck's constant
$C = 2.99792458 \times 10^{8}$ m/s, velocity of light
$\lambda_{moon} = 3.01624105596 \times 10^{-65}$ m, wavelength of moon

3)   Pershing's Energy Concept

$E_{moon} = \dfrac{\mathcal{A}_{moon} C^4}{G} = 6.58583788293 \times 10^{39}$ joules

$\mathcal{A}_{moon} = 5.44029768713 \times 10^{-5}$ m, Gravitational Structural Length of the Earth
$C^4 = 8.07760871307 \times 10^{33}$ m^4/s^4
$G = 6.67259 \times 10^{-11}$ m^3/ kg s^2 , Sir Isaac Newton's Gravitational Constant

$\mathbf{P_c} = \dfrac{C^4}{G} = 1.21056571932 \times 10^{44}$ kg m/s^2, D. Pershing's Energy Concept

Earth & Moon System

1) First Section

The Sidereal Month

A) 27 Days   x $\dfrac{86,400"}{1 \text{ Day}}$ = 2,332,800 sec

  7 Hours   x $\dfrac{3600"}{1 \text{ Hour}}$ =   25,200 sec

  43 Min   x $\dfrac{60"}{1 \text{ Min}}$ =   258 sec

  11.5 Sec ………………. = $\underline{\quad\quad 11.5 \text{ sec}}$

P_{moon} =   2360591.5 sec

B) P_{moon} = 27.3216608796 Days x $\dfrac{86,400 \text{ sec}}{1 \text{ Day}}$

  P_{moon} = 2,360,591.5 sec

C) a_{moon} = 3.844 x 10^8 meters
  $(a)^3$ = 5.6800235584 x 10^{25} m^3

D) $(P_{moon})^2$ = 5.57239222987 x 10^{12} sec^2

E) $\dfrac{(a)^3}{(P)^2} = k_{moon}$ = 1.01931510276 x 10^{13} m^3/s^2
  π^2 = 9.86960440109
  $4\pi^2$ = 39.4784176044

F) $(4\pi^2) \dfrac{(a)^3}{(P)^2}$ = 4.02409472972 x 10^{14} m^3/s^2

2) Second Section

A) $\dfrac{(4\pi^2)(a)^3}{(1-e^2)(P)^2}$ = 4.0362600579 x 10^{14} m^3/s^2
The eccentricity of the moon :
  e_m = 0.0549 & $(e_m)^2$ = 0.00301401
  $\dfrac{1}{1-(e_m)^2}$ = 1.00302312172

B) $\dfrac{(4\pi^2)\,(a)^3}{(1-(e_m)^2)\,(P)^2} = M_E + M_{moon}$

* Sir Isaac Newton's Gravitational Constant is $G = 6.67259 \times 10^{-11}$ m^3/kg s^2

Therefore, mass of earth & moon

$M_E + M_{moon} = 6.03078374322 \times 10^{24}$ kg

$81.3008130081 = M_{moon} = 7.32773268032 \times 10^{22}$ kg

$M_E = (M_E + M_{moon}) - M_{moon}$

C) $M_{moon} = 7.32773268032 \times 10^{22}$ kg, mass of the moon

$M_E = 5.957506244160 \times 10^{24}$ kg, mass of the earth

$\dfrac{M_E}{M_{moon}} = 80.3008130081$

$1 + \dfrac{M_E}{M_{moon}} = 81.3008130081 = \dfrac{1}{0.0123}$

Checks

A) $[(M_E + M_{moon})\,G] = 4.0362600579 \times 10^{14}$ m^3/ kg s^2

$(M_E + M_{moon}) = 6.04901553655 \times 10^{24}$ kg

$G = 6.67259 \times 10^{-11}$ m^3/ kg s^2, Sir Isaac Newton's Gravitational Constant

B) $v_{moon} = 1{,}024.7027002$ m/s, average orbital velocity of the moon

$(v_{moon})^2 = 1{,}050{,}015.6238$ m^2/s^2

$[(v_{moon})^2\,a_{moon}] = 4.0362600579 \times 10^{14}$ m^2/s^2

$a_{moon} = 3.844 \times 10^8$ m, distance to the moon

C) Final Conclusion

$\dfrac{(4\pi^2)\,(a)^3}{(1-e^2)\,(P)^2} = (M_E + M_{moon})\,G = (\mathcal{A}_E + \mathcal{A}_{moon})\,C^2 = (v_{moon})^2\,a_{moon}$

3) Third Section

A) $M_E G = 3.97519965897 \times 10^{14}$ m³/s²
$M_E = 5.957506244160 \times 10^{24}$ kg
$G = 6.67259 \times 10^{-11}$ m³/ kg s²

$\mathcal{A}_E = 4.42300612338 \times 10^{-3}$ m, Gravitational Structural Length of the Earth
$C^2 = 8.98755178737 \times 10^{16}$ m²/s²
$\mathcal{A}_E C^2 = 3.97519965899 \times 10^{14}$ m³/s²

Therefore, $M_E G = \mathcal{A}_E C^2$

4) Energy Concepts

A) First Energy Concept
$E_E = M_E C^2$, Einstein's Energy concept
$E_E = 5.3543395893 \times 10^{41}$ Joules

Data:
$M_E = 5.957506244160 \times 10^{24}$ kg
$C^2 = 8.98755178737 \times 10^{16}$ m²/s²

B) Second Energy Concept
$E_E = \dfrac{\mathcal{A}_E C^4}{G} = 5.3543395893 \times 10^{41}$ Joules, D. Pershing's Energy Concept

Data:
$\mathcal{A}_E = 4.42300612338 \times 10^{-3}$ m, Gravitational Structural Length of the Earth
$C^4 = 8.07760871307 \times 10^{33}$ m⁴/s⁴
$G = 6.67259 \times 10^{-11}$ m³/ kg s², Sir Isaac Newton's Gravitational Constant
$\mathbf{P}_c = \dfrac{C^4}{G} = 1.21056571932 \times 10^{44}$ kg m/s², D. Pershing's Energy Concept

C) Third Energy Concept – Max Planck's Energy Concept
$E_E = \hbar C / \lambda_E = \hbar \nu_E$
$E_E = 5.3543395893 \times 10^{41}$ Joules

Data:
$C = 2.99792458 \times 10^8$ m/s, velocity of light
$\lambda_E = 3.70997660478 \times 10^{-67}$ m, wavelength of earth
$\nu_E = C / \lambda_E = 8.08071020214 \times 10^{74}$ vibrations/s
$\hbar = 6.6260755 \times 10^{-34}$ Joules * s, max Planck's constant
$E_E = \hbar \nu_E = 5.3543395893 \times 10^{41}$ Joules

5) Density of the Earth

$$d_E = \frac{M_E}{(4\pi/3)\,(R_E)^3}$$

$d_E = 5.49717823388 \times 10^3 \text{ kg/m}^3$

$d_E \cong 5.5 \times 10^3 \text{ kg/m}^3$

Data:

$M_E = 5.957506244160 \times 10^{24}$ kg, mass of the earth

$R_E = 6,378,136$ m, radius of the earth

$(R_E)^3 = 4.06806188345 \times 10^{13} \text{ m}^3$

volume $_E = (4\pi/3)\,(R_E)^3 = 1.08685081532 \times 10^{21} \text{ m}^3$, volume of the earth

$4\pi = 12.5663706144$

$(4\pi/3) = 4.1887902048$

Ratio Concepts of the Earth & Moon

- Mass Ratio

$M_{Earth} = 5.957506244160 \times 10^{24}$ kg
$M_{moon} = 7.32773268032 \times 10^{22}$ kg

$$\frac{M_{Earth}}{M_{moon}} = 81.3008130081 = \frac{1}{.0123}$$

- Gravitational Structural Length Ratio

$\mathcal{A}_{Earth} = 4.42300612338 \times 10^{-3}$ m
$\mathcal{A}_{moon} = 5.44029753176 \times 10^{-5}$ m

$$\frac{\mathcal{A}_{Earth}}{\mathcal{A}_{moon}} = 81.3008130081 = \frac{1}{.0123}$$

- Equality Concept

$(M_{Earth} + M_{moon})\, G = 4.02409461478 \times 10^{14}$ m³/s²

$(\mathcal{A}_{Earth} + \mathcal{A}_{moon})\, C^2 = 4.02409461478 \times 10^{14}$ m³/s²

$[(v_{moon})^2\, a_{moon}] = 4.02409461478 \times 10^{14}$ m³/s²

$G = 6.67259 \times 10^{-11}$ m³/ kg s², Sir Isaac Newton's Gravitational Constant
$C^2 = 8.98755178737 \times 10^{16}$ m²/s²
$a_{moon} = 3.844 \times 10^8$ m, distance to the moon

Wavelength and GSL for the Earth & Moon

$\lambda_{moon} = 3.01624105596 \times 10^{-65}$ m, the wavelength of the moon

$\mathcal{A}_{moon} = 5.44029753176 \times 10^{-5}$ m, gravitational structural length of the moon

$(\lambda_{moon} \times \mathcal{A}_{moon}) = 1.64092492406 \times 10^{-69}$ m^2

$\lambda_{Earth} = 3.7099766048 \times 10^{-67}$ m, wavelength of the Earth

$\mathcal{A}_{Earth} = 4.42300612337 \times 10^{-3}$ m, gravitational structural length of the Earth

$(\lambda_{Earth} \times \mathcal{A}_{Earth}) = 1.64092492406 \times 10^{-69}$ m^2

Eccentricity of the Earth & Moon
& the Expansion Velocity of the Equinoxes

A) The Eccentricity of the Earth

$e_{earth} = 0.0167$

$(e_{earth})^2 = 0.00027889$

$[1 + (e_{earth})^2] = 1.00027889$

B) The Eccentricity of the Moon

$e_{moon} = 0.0549$

$(e_{moon})^2 = 0.000301401$

$[1 - (e_{moon})^2] = 0.99698599$

C) Ratio Concept

$$\frac{[1 + (e_{earth})^2]}{[1 - (e_{moon})^2]} = 1.00330285484$$

$$\frac{[1 + (e_{earth})^2]}{[1 - (e_{moon})^2]} \times 1.00336700337 = 1.00034284519$$

D) The Expansion Velocity of the Equinox

$v_{expansion\ velocity\ outward} = 3,389,941.43338$ m/s

$c = 2.99792458 \times 10^8$ m/s, velocity of light

$v/c = 0.01130762747$

$(v/c)^2 = 0.000127862439$

$[1 - (v/c)^2] = 0.999872137561$

$\sqrt{[1 - (v/c)^2]} = 0.999936066737$

$1/\sqrt{[1 - (v/c)^2]} = 1.00006393735$, Expansion Velocity Constant

E) Final Check

$(1/297) = 0.003367003367$

$1 + (1/297) = 1.00336700337$

$\phi = 49.8461538461$ arc sec

$\phi[1 + (1/297)] = 50.0139860141$ arc sec/yr

Earth-Moon System: Data of the Ellipse
"Kepler's Ellipse"

1) $v_{moon} = 1,023.14730277$ m/s, average orbital velocity of the moon

2) $v_p = v_{moon} (1 + e_{moon})$
$v_p = 1,079.31808969$ m/s
$e_{moon} = 0.0549$, eccentricity of the moon

3) $v_a = v_{moon} (1 - e)$
$v_a = 966.966515848$ m/s

$(1 - e) = 0.9451$

4) $v_{moon} = \dfrac{v_p + v_a}{2}$
$v_{moon} = 1,023.14730277$ m/s
$(v_p + v_a) = 2,046.29460554$ m/s

5) $a = 3.844 \times 10^8$ m, average distance for the Earth – Moon system
$a^2 = 1.4776336 \times 10^{17}$ m^2

6) $(a\,e) = 21,103,560$ m
$(a\,e)^2 = 4.45360244674 \times 10^{14}$ m^2

7) $[a^2 + (a\,e)^2] = 1.48208720245 \times 10^{17}$ m^2

8) $[a^2 + (a\,e)^2]^{\frac{1}{2}} = 384,978,856.88$ m

9)

$e^2 = 0.00301401$
$(1 + e^2) = 1.00301401$
$a^2(1 + e^2) = 1.48208720245 \times 10^{17}$ m^2

$(a\,e) = 21,103,560$ m
$(a\,e)^2 = 4.45360244674 \times 10^{14}$ m^2
$a^2 + (a\,e)^2 = 1.48208720245 \times 10^{17}$ m^2

10) Therefore:
$a^2(1 + e^2) = [a^2 + (a\,e)^2]$

11)

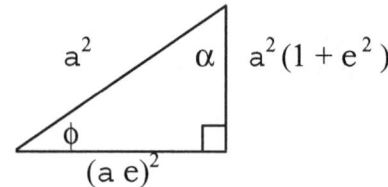

tan φ = $\frac{\text{opposite}}{\text{adjacent}}$

tan φ = $\frac{a^2(1+e^2)}{(a\,e)^2}$

tan φ = $\frac{1.48208720245 \times 10^{17}\ m^2}{4.45360244674 \times 10^{14}\ m^2}$

tan φ = 332.783902509

φ = $\tan^{-1}$ (332.783902509)

φ = 89.8278293913°

ε = 0.1721706087

φ + α = 89.8278293913° + 0.1721706087°

φ + α = 90°

12)

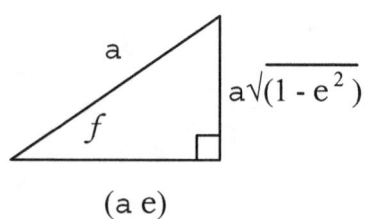

e = 0.0549
e^2 = 0.00301401
$(1-e^2)$ = 0.99698599
$(1+e^2)^{1/2}$ = 0.998491857753
$a(1-e^2)^{1/2}$ = 383,820,270.12 m
a = 3.844 × 10^8 m
(a e) = 21,103,560 m

tan φ = $\frac{a\sqrt{(1-e^2)}}{(a\,e)}$

tan φ = 18.1874655328

f = $\tan^{-1}\frac{a\sqrt{(1-e^2)}}{(a\,e)}$

f = 86.8528794438°

13) $a_p = a(1 - e)$, the perigee – the closest orbital point of the earth-moon system

$a_p = 363,296,440$ m

Data:
  $a = 3.844 \times 10^8$ m, average distance for the Earth – Moon system
  $e_{moon} = 0.0549$, eccentricity of the moon
  $(1 - e) = 0.9451$

14) $a_{ap} = a(1 + e)$, the apogree – the farthest distance between the Earth - Moon system
  $(1 + e) = 1.0549$
  $a_{ap} = 405,503,560$ m

15) $a_{ap} + a_p = 768,800,000$ m

$$a = \frac{a_{ap} + a_p}{2} = \frac{768,800,000 \text{ m}}{2} = 3.844 \times 10^8 \text{ m}$$

Escape Velocity of the Moon

$$v_{es} = e \sqrt{(2 \,\mathcal{A}_{moon}) / \mathcal{R}_{moon})} = 2,372.04075714 \text{ m/s}$$

$$x \,(1 \text{ mile}/ 1.6093 \times 10^3 \text{ m}) = 1.4739580918 \text{ miles/s}$$

$$v_{es} \approx 1.50 \text{ miles/s}$$

Data:

$\mathcal{A}_{moon} = 5.44029768713 \times 10^{-5} \text{ m}$
$\mathcal{R}_{moon} = 1,738,000 \text{ m}$
$(\mathcal{A}_{moon} / \mathcal{R}_{moon}) = 3.13020580388 \times 10^{-11} \text{ rad}$

Westward Precession of the Equinoxes

- First Concept of the Precession of the Equinoxes

$$\Phi = \left[\frac{(1 + e_{moon})^2}{(1 - e_E)^2} \right] \left[\frac{360° \times (60'/1°) \times (60''/60')}{26,000 \ \text{years}} \right]$$

$$\Phi = (1.00329381861) \left[\frac{1,296,000 \ \text{arc sec}}{26,000 \ \text{years}} \right]$$

$$\Phi = 50.0103380354 \ \text{sec of arc/yr} \qquad \text{D. Pershing's Concept}$$

Data:

the eccentricity of the moon
$e_{moon} = 0.0549$
$e_{moon}^2 = 0.00301401$
$[1 + e_{moon}^2] = 1.00301401$

the eccentricity of the earth
$e_E = 0.0167$
$e_E^2 = 0.00027889$
$[1 - e_E^2] = 0.99972111$

$$\frac{(1 + e_{moon})^2}{(1 - e_E)^2} = 1.00329381861$$

$$\frac{1,296,000 \ \text{arc sec}}{26,000 \ \text{years}} = 49.8461538462 \ \text{sec of arc/yr}$$

- Second Concept of the Precession of the Equinoxes

$$\Phi = 50.0103380354 \text{ sec of arc/yr}$$

Data:

* See Hayford's work in <u>The VNR Concise Encyclopedia of Mathematics</u> by Von Nostrand Reinhold Company

The flattening of the earth = 1/297 (as the earth is an ellipsoid)

$$1/297 = 0.003367003367$$

$$1 + 1/297 = 1.003367003367$$

$$(1 + 1/297) \, (49.8461538462 \text{ sec of arc/yr}) = 49.842518109 \text{ sec of arc/yr}$$

Ratio of 2nd concept/1st concept = 1.0000729449

$$49.842518109 \text{ sec of arc/yr} \times 1.003367003367 = 50.0103380354 \text{ sec of arc/yr} = \Phi$$

The Sun

M_{sun} = 1.98891931841 x 10^{30} kg

$\mathcal{A}_{sun}$ = 1,476.62494401 m

c = 2.99792458 x 10^8 m/s

t_{sun} = 4.92549063396 x 10^{-6} sec, Time for light to travel the curvature length of the bending of space. The curvature length is the Gravitational Structural Length.

$\mathcal{A}_{sun}$ t_{sun} C

t_{sun} = (1,476.62494401 m) / (2.99792458 x 10^8 m/s)
t_{sun} = 4.92549063396 x 10^{-6} sec

The Earth and Sun System

1)   The sidereal period of the earth:

365 days	x	86400 s/day	=	31,536,000 s
6 hrs	x	3600 s/hr	=	21,600 s
9 min	x	60 s/min	=	540 s
9.54 sec	x		=	9.54 s
		P_{earth}	=	31,558,149.54 s

2) P_{earth} = 365.256360417 days x 86400s/day
P_{earth} = 31558149.54 s
$(P_{earth})^2$ = 9.95916802389 x 10^{14} s^2

3) a_{sun} = 1.4959787 x 10^{11} m, "one astronomical unit"
$(a_{sun})^3$ = 3.34792892881 x 10^{33} m^3

4) $\dfrac{(a_{sun})^3}{(P_{earth})^2}$ = 3.36165523142 x 10^{18} m^3/s^2

5) $(4\pi^2)\dfrac{(a_{sun})^3}{(P_{earth})^2}$ = 1.32712829068 x 10^{20} m^3/s^2
$4\pi^2$ = 39.4784176044

$$\left[\dfrac{(4\pi^2)(a_{sun})^3}{(P_{earth})^2}\right] \bigg/ \; G = (M_{sun} + M_{earth})$$

$$\left[\dfrac{(4\pi^2)(a_{sun})^3}{(P_{earth})^2}\right] = 1.98892527591 \times 10^{30} \text{ kg}$$

G = 6.67259 x 10^{-11} m^3/s^2, Sir Isaac Newton's Gravitational Constant

6) $M_{sun} = (M_{sun} + M_{earth}) - (M_{earth})$
M_{sun} = (1.98892527591 x 10^{30} kg) – (5.97461264545 x 10^{24} kg)
M_{sun} = 1.9889193184 x 10^{30} kg

7) $(M_{sun} G)$ = 1.32712431548 x 10^{20} m^3/s^2
A_{sun} = 1,476.62494401 m, gravitational structural length of the sun
C = 2.99792458 x 10^8 m/s
C^2 = 8.98755178737 x 10^{16} m^2/s^2
$(A_{sun} C^2)$ = 1.32712431548 x 10^{20} m3/s^2
Therefore, $(A_{sun} C^2) = (M_{sun} G)$

Gravitational Structural Lengths
Concepts of the Sun and Earth

A) $\dfrac{(4\pi^2)\,(a)^3}{(P)^2} = 4.03625856541 \times 10^{14}\ m^3/s^2$

$(\mathcal{A}_E + \mathcal{A}_{moon}) = 4.49094331905 \times 10^{-3}\ m$, Gravitational Structural Lengths of
the Sun and Earth

$C = 2.99792458 \times 10^8\ m/s$

$C^2 = 8.98755178737 \times 10^{16}\ m^2/s^2$

B) $(\mathcal{A}_E + \mathcal{A}_{moon})\,C^2 = 4.03625856541 \times 10^{14}\ m^3/s^2$

C) Average Orbital Velocity of Earth

$v_E = \dfrac{2\pi a}{P} = 29{,}784.7355586\ m/s$

$2\pi = 6.28318530718$

$a_s = 1\ AU = 1.4959787 \times 10^{11}\ meters$

$P_E = 31558149.54\ s$, sidereal period of the earth

$v_E = 29{,}784.7355586\ m/s$, average orbital velocity of the earth

$(v_E)^2 = 887130472.296\ m^2/s^2$

$(v_E)^2\,a_s = (887130472.296\ m^2/s^2)\,(1.4959787 \times 10^{11}\ m)$

$(v_E)^2\,a_s = 1.32712829068 \times 10^{20}\ m^3/s^2$

Deflection of Light by the Sun

$$D = \frac{4\,(\mathcal{A}_{sun})}{(\mathcal{R}_{sun})} \times \frac{(1{,}296{,}000 \text{ sec})}{2\pi}$$

$$D = \frac{4(1{,}476.62494401 \text{ m})}{6.96 \times 10^8 \text{ m}} \times \frac{(1296000 \text{ sec})}{2\pi\text{rad}}$$

$$D = 1.75043539066 \text{ sec}$$

Deflection of Light by the Sun and Earth

$$D_{sun} = \left[\frac{4(A_{sun} + A_{earth})}{R_{sun}}\right] \times \left[\frac{1,296,000 \text{ s}}{2\pi}\right] = 1.75044063383" \text{ of arc}$$

Data:

$2\pi = 6.28318530718$

$(A_{sun} + A_{earth}) = 1,476.62936702$ m,

Gravitational Structural Length of the Sun & Earth

$R_{sun} = 6.96 \times 10^8$ m, radius of the sun

$4(A_{sun} + A_{earth}) = 5,906.5174808$ m

$$\left[\frac{4(A_{sun} + A_{earth})}{R_{sun}}\right] = 8.48637567253 \times 10^{-6}$$

$(360° \times 60'/1° \times 60"/1') = 1,296,000$ sec

$$\left[\frac{1,296,000 \text{ sec}}{2\pi}\right] = 206,264.806247$$

Deflection of Light by the Sun, Earth & Moon at Total Eclipse

$$D = \frac{4\,(\mathcal{A}_{sun} + \mathcal{A}_{earth} + \mathcal{A}_{moon})}{(\mathcal{R}_{sun} + \mathcal{R}_{earth} + \mathcal{R}_{moon})} \cdot \frac{(1,296,000 \text{ sec})}{2\pi}$$

$$D = 1.7302638922 \text{ sec of arc}$$

Data:

- $M_{sun} = 1.9889193184 \times 10^{30}$ kg, mass of sun
- $M_{earth} = 5.957506244160 \times 10^{24}$ kg, mass of earth
- $M_{moon} = 7.32773268032 \times 10^{22}$ kg, mass of moon
  $G = 6.67259 \times 10^{-11}$ m^3/kg s^2 , Sir Isaac Newton's Gravitational Constant

$$M_{sun}\, G = \mathcal{A}_{sun}\, C^2$$

A) $M_{sun}\, G = 1.32712431548 \times 10^{20}$ m^3/s^2
       $C^2 = 8.98755178737 \times 10^{16}$ m^2/s^2
       $\mathcal{A}_{sun} = 1,476.62494401$ meters, gravitational structural length of the sun

B) $M_{earth}\, G = 3.97519965897 \times 10^{14}$ m^3/s^2
       $\mathcal{A}_{earth} = \dfrac{M_{earth}\, G}{C^2} = 4.42300612338 \times 10^{-3}$ m

C) $M_{moon}\, G = 4.88949558054 \times 10^{12}$ m^3/s^2
       $\mathcal{A}_{moon} = \dfrac{M_{moon}\, G}{C^2} = 5.44029753176 \times 10^{-5}$ m

D) $\mathcal{A}_{sun} + \mathcal{A}_{earth} + \mathcal{A}_{moon} = 1,476.62942142$ m

E) $\mathcal{R}_{sun} = 6.96 \times 10^8$ m
       $\mathcal{R}_{earth} = 6,378,136$ m
       $\mathcal{R}_{moon} = 1,738,000$ m
       $\mathcal{R}_{sun} + \mathcal{R}_{earth} + \mathcal{R}_{moon} = 704,116,136$ m

F) $\dfrac{(\mathcal{A}_{sun} + \mathcal{A}_{earth} + \mathcal{A}_{moon})}{(\mathcal{R}_{sun} + \mathcal{R}_{earth} + \mathcal{R}_{moon})} = 2.09713901716 \times 10^{-6}$ rad

G) $\dfrac{4(\mathcal{A}_{sun} + \mathcal{A}_{earth} + \mathcal{A}_{moon})}{(\mathcal{R}_{sun} + \mathcal{R}_{earth} + \mathcal{R}_{moon})} = 8.388555606864 \times 10^{-6}$ rad

H) $(360°)\,(60 \text{ min}/°)\,(60 \text{ sec/min}) = 1,296,000$ sec
       $2\pi = 6.28318530718$

I) $\dfrac{(1,296,000 \text{ sec})}{2\pi} = 206,264.806247$ sec of arc

Therefore:

$$D = \frac{4\,(\mathcal{A}_{sun} + \mathcal{A}_{earth} + \mathcal{A}_{moon})}{(\mathcal{R}_{sun} + \mathcal{R}_{earth} + \mathcal{R}_{moon})} \cdot \frac{(1,296,000 \text{ sec})}{2\pi}$$

$D = 1.7302638922$ sec of arc, deflection of light at a total eclipse

Pershing's Area Constant for the Sun

$\mathcal{A}_{Sun} \; \lambda_{Sun} = (\hbar \, G) / \, C^3$

$\hbar = 6.6260755 \times 10^{-34} \text{ kg m}^2/\text{s}$

$G = 6.67259 \times 10^{-11} \text{ m}^3/ \text{kg s}^2$

$C = 2.99792458 \times 10^8 \text{ m/s}$

$C^3 = 2.69440024174 \times 10^{25} \text{ m}^3/\text{s}^3$

$\mathcal{A}_{Sun} \; \lambda_{Sun} = (\hbar \, G) / \, C^3 = 1.64092492406 \times 10^{-69} \text{ m}^2$

Therefore, $\mathcal{A}_{Earth} \; \lambda_{Earth} = \mathcal{A}_{Sun} \; \lambda_{Sun} = 1.64092492406 \times 10^{-69} \text{ m}^2$

D. Pershing's Area Constant for the Planets & Sun

1) Using the mass of the earth:

$M_E = 5.957506244160 \times 10^{24}$ kg

The total energy of the earth $E_E = M_E C^2$

$C = 2.99792458 \times 10^8$ m/s & $C^2 = 8.98744178737 \times 10^{16}$ m^2/s^2
Therefore, $E_E = 5.3543395893 \times 10^{41}$ kg m^2/s^2

2) $E_E = h f_E$ $\qquad\qquad$ $E_E = M_E C^2$

$h f_E = M_E C^2$ $\qquad\qquad$ Therefore, $f_E = \dfrac{M_E C^2}{h}$

$h = 6.6260755 \times 10^{-34}$ kg m^2/s, Planck's constant

$f_E = 8.08071020214 \times 10^{74}$ vibrations/s, vibration of the earth
$f_E = \dfrac{C}{\lambda}$ so $\lambda_E = \dfrac{C}{f_E}$

$\lambda_E = \dfrac{2.99792458 \times 10^8 \text{ m/s}}{8.08071020214 \times 10^{74} \text{ vibrations/s}}$

$\lambda_E = 3.70997660479 \times 10^{-67}$ m

3) The gravitational structural length of the earth: the distortion (or dip) in space

$M_E G = 3.97519965897 \times 10^{14}$ m^3/s$^2 = A_E C^2$
$G = 6.67259 \times 10^{-11}$ m^3/kg s^2
$A_E = \dfrac{M_E G}{C^2} = 4.42300612338 \times 10^{-3}$ m

4) $A_E \lambda_E = 1.64092492406 \times 10^{-69}$ m^2

Concepts of $MG = AC^2$: Mercury

$M_{Mercury} = 3.3022 \times 10^{23}$ kg

$G_{Earth} = 6.67259 \times 10^{-11}$ m^3/kg s^2

$(M_{Mercury})(G_{Earth}) = 2.2034226698 \times 10^{13}$ m^3/s^2

$A_{Mercury} = 2.45163835706 \times 10^{-4}$ m, Gravitational Structural Length of Mercury

$C^2 = 8.98755178737 \times 10^{16}$ m^2/s^2

$(A_{Mercury})C^2 = 2.2034226698 \times 10^{13}$ m^3/s^2

Therefore, $(M_{Mercury})(G_{Earth}) = (A_{Mercury})C^2$

Perihelion Anomaly of Mercury

Einstein's Method

$$\Delta \omega_{\text{Mercury}} = \left(\frac{3}{(1-e^2)} \right) \left(\frac{v_{\text{Mercury}}}{C} \right)^2 \left(\frac{1,296,000"}{\text{century}} \right) \left(\frac{100 \text{ years}}{\text{Tropical yr of Mercury}} \right)$$

Pershing's Method

$$\Delta \omega_{\text{Mercury}} = \left(\frac{3}{(1-e^2)} \right) \left(\frac{(\mathcal{A}_{\text{Sun}} + \mathcal{A}_{\text{Mercury}})}{a_{\text{Mercury}}} \right) \left(\frac{v_{\text{Mercury}}}{C} \right)^2 \left(\frac{1,296,000"}{\text{century}} \right) \left(\frac{100 \text{ years}}{\text{Tropical yr}} \right)$$

Data:

$\Delta \omega_{\text{Mercury}} = 42.9801297691$ sec/century

$v_{\text{Mercury}} = 47,871.945305$ m/s

$(\mathcal{A}_{\text{Sun}} + \mathcal{A}_{\text{Mercury}}) = 1,476.62860941$ m

Tropical yr of Mercury $= P_{\text{Mercury}} / P_{\text{Earth}} = 0.240844539708$ yr

$P_{\text{Mercury}} = 87.97$ days

$P_{\text{Earth}} = 365.256360417$ days

$e = 0.2056$

Perturbation of Mercury

First Concept: Average Distance to Mercury

Data:
$$a_{mer} = 0.387100660269 \text{ AU} \times \frac{1.4959787 \times 10^{11} \text{ m}}{1 \text{ AU}}$$

$$a_{mer} = 57,909,434,251.8 \text{ m}$$

$$(a_{mer})^3 = 1.94199436901 \times 10^{32} \text{ m}^3$$

Second Concept: Sidereal Period of Mercury

Data:
$$P_{mer} = 87.97 \text{ days} \times (86400 \text{ s} / 1 \text{ day})$$

$$P_{mer} = 7,600,608 \text{ s}$$

$$(P_{mer})^2 = 5.77692419697 \times 10^{13} \text{ s}^2$$

Correction Factor for Perturbation of Mercury

- $\Delta\omega_{mer}$ = 42.9805189898 sec of arc/century
  x 1.00047912748 "correction factor"

- $\Delta\omega_{mer}$ = 43.00111121376 sec of arc/century

Data:

$$\frac{(v_{mer})}{(C)} = \frac{(47,871.9565255 \text{ m/s })}{(2.99792458 \times 10^8 \text{ m/s })}$$

$$\frac{(v_{mer})}{(C)} = 1.5963658638 \times 10^{-4} \text{ rad}$$

$$\left[1 + \frac{(v_{mer})}{(C)} \right] = 1.00015968366 \text{ rad}$$

$$\left[1 + \frac{(v_{mer})}{(C)} \right]^3 = 1.00047912748 \text{ rad}^3$$

Concepts of MG = AC^2 : Venus

M $_{\text{Venus}}$ = 4.869 x 10^{24} kg
G $_{\text{Earth}}$ = 6.67259 x 10^{-11} m^3/kg s^2
 (M $_{\text{Venus}}$)(G $_{\text{Earth}}$) = 3.248884071 x 10^{14} m^3/s^2

$\mathcal{A}$ $_{\text{Venus}}$ = 3.61487104371 x 10^{-3} m, Gravitational Structural Length of Venus
C ^{2} = 8.98755178737 x 10^{16} m^2/s^2
 ($\mathcal{A}$ $_{\text{Venus}}$)C ^{2} = 3.248884071 x 10^{14} m^3/s^2

Therefore, (M $_{\text{Venus}}$)(G $_{\text{Earth}}$) = ($\mathcal{A}$ $_{\text{Venus}}$)C ^{2}

Venus' Centripetal Force

1)   First Concept of Centripetal Force

$$F_{Venus} = \frac{M_{Venus} (v_{Venus})^2}{a_{Venus}}$$ Average centripetal force

$$F_{Venus} = 5.45449544933 \times 10^{28} \text{ kg m/s}^2$$

Data:

$$M_{Venus} = 4.81247354403 \times 10^{30} \text{ kg}$$

$$v_{Venus} = 35{,}020.6918462 \text{ m/s}$$

$$(v_{Venus})^2 = 1{,}226{,}448{,}857.39 \text{ m}^2/\text{s}^2$$

$$a_{Venus} = 108{,}208{,}957{,}806 \text{ m, distance to the sun}$$

2)   Second Concept of Centripetal Force

$$F_{Venus} = \frac{M_{Venus} [(\mathcal{A}_{Venus} + \mathcal{A}_{moon})C^2]}{(a_{Venus})^2}$$ Average centripetal force

$$F_{Venus} = 5.45449544933 \times 10^{28} \text{ kg m/s}^2$$

* Uses a "double gravitational structural length" concept to replace Newton's "double mass concept."

Data:

$$(a_{Venus})^2 = 1.17091785495 \times 10^{22} \text{ m}^2$$

$$M_{Venus} = 4.81247354403 \times 10^{30} \text{ kg}$$

$$(\mathcal{A}_{Venus} + \mathcal{A}_{moon}) = 1{,}476.62851687 \text{ m}$$

$$C^2 = 8.98755178737 \times 10^{16} \text{ m}^2/\text{s}^2$$

$$(\mathcal{A}_{Venus} + \mathcal{A}_{moon})C^2 = 1.32712752661 \times 10^{20} \text{ m}^3/\text{s}^2$$

The Perturbation of Venus

$$\Delta \varpi_{venus} = \left[\frac{3}{(1-e^2)} \right] \left[\frac{v}{C} \right]^2 \left[\frac{P_E}{P_{venus}} \right] (1296000") \left[\frac{100 \text{ years}}{\text{century}} \right]$$

$$\Delta \varpi_{venus} = 8.6248220979"$$

Data:

$$\left[\frac{P_E}{P_{venus}} \right] = \left[\frac{365.256365741 \text{ days}}{224.7 \text{ days}} \right] \Big/ 1 \text{ year}$$

$$\left[\frac{P_E}{P_{venus}} \right] = 1.6255289975 \text{ rad} \Big/ 1 \text{ year}$$

$$\left[\frac{v}{C} \right] = \frac{35020.734194 \text{ m/s}}{2.99792458 \times 10^8 \text{ m/s}}$$

$$\left[\frac{v}{C} \right] = 1.16816595146 \times 10^{-4} \text{ rad}$$

$$\left[\frac{v}{C} \right]^2 = 1.364116859 \times 10^{-8} \text{ rad}$$

$$(1 - (e_{venus})^2) = 1 - (0.0068)^2 = 0.99995376$$

$$\left[\frac{3}{(1-e^2)} \right] \left[\frac{v}{C} \right]^2 = 4.0942436539 \times 10^{-8} \text{ rad}$$

The Escape Velocity from Venus

Basic Equation:

$$v_{escape} = e \left[\frac{(2\,M_{Venus})\,G}{(R_{Venus})} \right]^{\frac{1}{2}}$$

$v_{escape} = 10{,}361.8245316$ m/s

$v_{escape} = 10.3618245316$ km/s

$M_{Venus} = 4.869 \times 10^{24}$ kg
$G = 6.67259 \times 10^{-11}$ m^3/kgs^2
$R_{Venus} = 6{,}051{,}900$ m

Pershing's Concept for Escape Velocity:

$$v_{escape} = \sqrt{(2\,A_{Venus})\,/\,R_{Venus}}$$

$v_{escape} = 10.3618245316$ km/s

Perihelion Anomaly of Venus

Einstein's Method

$$\Delta \omega_{\text{Venus}} = \left(\frac{3}{(1-e^2)} \right) \left(\frac{v_{\text{Venus}}}{C} \right)^2 \left(\frac{1,296,000"}{\text{century}} \right) \left(\frac{100 \text{ years}}{\text{Tropical yr of Venus}} \right)$$

Pershing's Method

$$\Delta \omega_{\text{Venus}} = \left(\frac{3}{(1-e^2)} \right) \left(\frac{(\mathcal{A}_{\text{Sun}} + \mathcal{A}_{\text{Venus}})}{a_{\text{Venus}}} \right) \left(\frac{v_{\text{Venus}}}{C} \right)^2 \left(\frac{1,296,000"}{\text{century}} \right) \left(\frac{100 \text{ years}}{\text{Tropical yr}} \right)$$

Data:

$\Delta \omega_{\text{Venus}} = 8.62521410285$ sec/century

$a_{\text{Venus}} = 108,208,703,229$ m

$(\mathcal{A}_{\text{Sun}} + \mathcal{A}_{\text{Venus}}) = 1,476.62860941$ m

Tropical yr of Venus $= P_{\text{Venus}} / P_{\text{Earth}} = 0.6151848369653$ yr

$e = 0.0068$

Concepts of MG = AC^2 : Mars

M $_{Mars}$ = 6.4191 x 10^{23} kg
G $_{Earth}$ = 6.67259 x 10^{-11} m^3/kg s^2
(M $_{Mars}$)(G $_{Earth}$) = 4.2832022469 x 10^{13} m^3/s^2

A $_{Mars}$ = 4.7657052201 x 10^{-4} m, Gravitational Structural Length of Mars
C ^{2} = 8.98755178737 x 10^{16} m^2/s^2
(A $_{Mars}$)C ^{2} = 4.2832022469 x 10^{13} m^3/s^2

Therefore, (M $_{Mars}$)(G $_{Earth}$) = (A $_{Mars}$)C ^{2}

Perturbation of Mars

First Method

$$\Delta \omega_{Mars} = 3 \left(\frac{v_{Mars}}{C} \right)^2 \left(\frac{1,296,000 \text{ sec}}{1 \text{ Tropical yr of Mars}} \right) \left(\frac{100 \text{ yr}}{\text{century}} \right)$$

$\Delta \omega_{Mars} = 1.33914642831$ sec/century

Data:

$v_{Mars} = 24,129.2930453$ m/s, orbital velocity of Mars

$C = 2.99792458 \times 10^8$ m/s, velocity of light

$(v_{Mars} / C) = 8.04866580243 \times 10^{-5}$ rad

$(v_{Mars} / C)^2 = 6.47810211992 \times 10^{-9}$ rad

$e_{Mars} = 0.0167$, eccentricity of the Mars

Second Method

$$\Delta \omega_{Mars} = 3 \left(\frac{(A_{Sun} + A_{Mars})}{a_{Mars}} \right) \left(\frac{1,296,000 \text{ sec}}{1 \text{ Tropical yr of Mars}} \right) \left(\frac{100 \text{ yr}}{\text{century}} \right)$$

$\Delta \omega_{Mars} = 1.33914642831$ sec/century

Data:

$(A_{Sun} + A_{Mars}) = 1,476.62547114$ m

$a_{Mars} = 227,941,061,101$ m

66

Mars' Motion of the Perihelion

$$\Delta \omega_{Mars} = \left[\frac{3}{(1-e^2)} \right] \left[\frac{v_{Mars}}{C} \right]^2 \left[\frac{P_{Earth}}{P_{Mars}} \right] \left[\frac{1,296,000"}{1 \text{ yr}} \right] \left[\frac{100 \text{ yr}}{\text{century}} \right]$$

$\Delta \omega_{Mars} = 1.35090491652$ sec

Data:

$v_{Mars} = 24,129.2930453$ m/s

$C = 2.99792458 \times 10^8$ m/s, velocity of light

$(v_{Mars} / C) = 8.04866580243 \times 10^{-5}$ rad

$(v_{Mars} / C)^2 = 6.47810211992 \times 10^{-9}$ rad

$e_{Mars} = 0.0167$, eccentricity of the Mars

$(1-e^2) = 0.99972111$

$$\left(\frac{P_{Earth}}{P_{Mars}} \right) = \frac{365.256365741 \text{ days}}{689.98 \text{ days}} = 0.529372395926$$

Concepts of $MG = AC^2$: Jupiter

$M_{\text{Jupiter}} = 1.8988 \times 10^{27}$ kg
$G_{\text{Earth}} = 6.67259 \times 10^{-11}$ m^3/kg s^2
$(M_{\text{Jupiter}})(G_{\text{Earth}}) = 1.2669913892 \times 10^{17}$ m^3/s^2

$\mathcal{A}_{\text{Jupiter}} = 1.40971804$ m, Gravitational Structural Length of Jupiter
$C^2 = 8.98755178737 \times 10^{16}$ m^2/s^2
$(\mathcal{A}_{\text{Jupiter}})C^2 = 1.2669913892 \times 10^{17}$ m^3/s^2

Therefore, $(M_{\text{Jupiter}})(G_{\text{Earth}}) = (\mathcal{A}_{\text{Jupiter}})C^2$

Jupiter

1) $a = 5.20280378653$ AU x $\underline{\frac{1.4959787 \times 10^{11} \text{ m}}{1 \text{ AU}}}$

   $a = 778{,}328{,}364{,}493$ m

   $a^3 = 4.71507465016 \times 10^{35}$ m^3

2) $P = 4{,}332.588$ days x $(86400$ s $/ 1$ day$)$

   $P = 374{,}335{,}603.2$ s

   $P^2 = 1.40127143823 \times 10^{17}$ s^2

3) $k = (a^3 / P^2) = 3.36485460384 \times 10^{18}$ m^3/s^2

   $(4\pi^2)\,(a^3 / P^2) = 1.32839135229 \times 10^{20}$ m^3/s^2

4) $(M_{Sun} + M_{Jupiter})\,G = (4\pi^2)\,(a^3 / P^2)$

   $G = 6.67259 \times 10^{-11}$ m^3/kgs^2, Newton's Gravitational Constant

5) $M_{Sun} = 1.98891938647 \times 10^{30}$ kg, mass of the sun

   $M_{Jupiter} = 1.89880001 \times 10^{27}$ kg, mass of Jupiter

   $(M_{Sun} + M_{Jupiter}) = 1.99081818648 \times 10^{30}$ kg

6) $(\mathcal{A}_{Sun} + \mathcal{A}_{Jupiter}) = 1{,}478.03471259$ m

   $C = 2.99792458 \times 10^{8}$ m/s

   $C^2 = 8.98755178737 \times 10^{16}$ m^2/s^2

7) To find average orbital velocity,

   $v^2\,a = (\mathcal{A}_{Sun} + \mathcal{A}_{Jupiter})\,C^2$

   $v = 13.0641630189$ km/s, average orbital velocity of Jupiter

8) Another method to find average orbital velocity, $v = (2\pi a) / P$

$(4\pi^2)\,(a^3 / P^2) = v^2\,a = (\mathcal{A}_{Sun} + \mathcal{A}_{Jupiter})\,C^2 = (M_{Sun} + M_{Jupiter})\,G$

The Perihelion Anomaly of Jupiter

$$\Delta \omega_{\text{Jupiter}} = \left(\frac{6\pi}{(1-e^2)} \right) \left(\frac{v_{\text{Jupiter}}}{C} \right)^2 \left(\frac{1,296,000"}{2\pi\text{rad}} \right) \left(\frac{100 \text{ /century}}{\text{tropical yr}} \right)$$

$$\Delta \omega_{\text{Jupiter}} = 0.063901150206 \text{ sec}$$

Data:

$v_{\text{Jupiter}} = 13,064.1630188$ m/s

$C = 2.99792458 \times 10^8$ m/s, velocity of light

$(v_{\text{Jupiter}} / C) = 4.35773571689 \times 10^{-5}$ rad

$(v_{\text{Jupiter}} / C)^2 = 1.89898605783 \times 10^{-9}$ rad

$e_{\text{Jupiter}} = 0.04837$, eccentricity of the Jupiter

Tropical yr of Jupiter $= 11.8617731097$ yr

Concepts of MG = AC^2 : Saturn

M $_{Saturn}$ = 5.685 x 10^{26} kg
G $_{Earth}$ = 6.67259 x 10^{-11} m^3/kg s^2
 (M $_{Saturn}$)(G $_{Earth}$) = 3.793367415 x 10^{16} m^3/s^2

$\mathcal{A}$ $_{Saturn}$ = 0.422069046693 m, Gravitational Structural Length of Saturn
C ^{2} = 8.98755178737 x 10^{16} m^2/s^2
 ($\mathcal{A}$ $_{Saturn}$)C ^{2} = 3.793367415 x 10^{16} m^3/s^2

Therefore, (M $_{Saturn}$)(G $_{Earth}$) = ($\mathcal{A}$ $_{Saturn}$)C ^{2}

71

Concepts of $MG = AC^2$: Uranus

$M_{Uranus} = 1.919139 \times 10^{25}$ kg

$G_{Earth} = 6.67259 \times 10^{-11}$ m^3/kg s^2

$(M_{Uranus})(G_{Earth}) = 1.28056277 \times 10^{15}$ m^3/s^2

$A_{Uranus} = 1.42481823782 \times 10^{-2}$ m, Gravitational Structural Length of Uranus

$C^2 = 8.98755178737 \times 10^{16}$ m^2/s^2

$(A_{Uranus})C^2 = 1.28056277 \times 10^{15}$ m^3/s^2

Therefore, $(M_{Uranus})(G_{Earth}) = (A_{Uranus})C^2$

Concepts of $MG = AC^2$: Neptune

$M_{\text{Neptune}} = 1.0278 \times 10^{26} \text{ kg}$

$G_{\text{Earth}} = 6.67259 \times 10^{-11} \text{ m}^3/\text{kg s}^2$

$(M_{\text{Neptune}})(G_{\text{Earth}}) = 6.858088002 \times 10^{15} \text{ m}^3/\text{s}^2$

$\mathcal{A}_{\text{Neptune}} = 7.63065199984 \times 10^{-2} \text{ m}$, Gravitational Structural Length of Neptune

$C^2 = 8.98755178737 \times 10^{16} \text{ m}^2/\text{s}^2$

$(\mathcal{A}_{\text{Neptune}})C^2 = 6.858088002 \times 10^{15} \text{ m}^3/\text{s}^2$

Therefore, $(M_{\text{Neptune}})(G_{\text{Earth}}) = (\mathcal{A}_{\text{Neptune}})C^2$

Concepts of MG = AC^2 : Pluto

M $_{\text{Pluto}}$ = 1.5 x 10^{22} kg

G $_{\text{Earth}}$ = 6.67259 x 10^{-11} m^3/kg s^2

(M $_{\text{Pluto}}$)(G $_{\text{Earth}}$) = 1.0008885 x 10^{12} m^3/s^2

A $_{\text{Pluto}}$ = 1.11363864563 x 10^{-5} m, Gravitational Structural Length of Pluto

C ^{2} = 8.98755178737 x 10^{16} m^2/s^2

(A $_{\text{Pluto}}$)C ^{2} = 1.0008885 x 10^{12} m^3/s^2

Therefore, (M $_{\text{Pluto}}$)(G $_{\text{Earth}}$) = (A $_{\text{Pluto}}$)C ^{2}

Sidereal Periods of the Planets

Mercury	P = 87.9693 days = 7,600,547.52 sec
Venus	P = 224.7008 days = 19,414,149.12 sec
Earth	P = 365.256354167 days = 31,558,149 sec
Mars	P = 686.9797 days = 59,335,016.08 sec
Jupiter	P = 4,332.588 days = 374,335,603.2 sec
Saturn	P = 10,759.201 days = 929,594,996.4 sec
Uranus	P = 30,685.93 days = 2,651,264,352 sec
Neptune	P = 60,187.64 days = 5,200,212,096 sec
Pluto	P = 90,737.2 days = 7,839,694,080 sec

Sidereal Periods of the Planets Squared

Mercury	$P^2 = 5.77683226038 \times 10^{13} \ s^2$
Venus	$P^2 = 3.76909186054 \times 10^{14} \ s^2$
Earth	$P^2 = 9.95916768306 \times 10^{14} \ s^2$
Mars	$P^2 = 3.52302149516 \times 10^{15} \ s^2$
Jupiter	$P^2 = 1.40127143823 \times 10^{17} \ s^2$
Saturn	$P^2 = 8.64146801556 \times 10^{17} \ s^2$
Uranus	$P^2 = 7.02920266419 \times 10^{18} \ s^2$
Neptune	$P^2 = 2.70422058434 \times 10^{19} \ s^2$
Pluto	$P^2 = 6.1460803268 \times 10^{19} \ s^2$

Proof for the Rydberg Constant

R_∞ = 10,973,731.6224 /meter, Rydberg constant

v_e = 2,187,691.43195 m/s, velocity of electron in 1st shell of Hydrogen atom

C = 2.99792458 x 10^8 m/s, velocity of light

$\dfrac{(v_e)}{(C)}$ = 7.29735313071 x 10^{-3}

4π = 12.5663706144

a_o = 5.29177249 x 10^{-11} m, Bohr's radius

$$\left(\frac{\dfrac{(v_e)}{(C)}}{4\pi\, a_o} \right) = 10,973,731.6224 \text{ /meter} = R_\infty$$

Formula can be rearranged to find the Velocity of the Electron

$v_e = (C) * (R_\infty) * (a_o) * (4\pi)$

v_e = (2.99792458 x 10^8 m/s) * (10,973,731.6224 /meter) * (5.29177249 x 10^{-11} m) * (12.5663706144)

v_e = 2,187,691.43195 m/s

Gravitational Structural Length of the Universe

A) Mass of the Universe

$M_{uni} = 2.23587060896 \times 10^{52}$ kg, mass of the universe
$G = 6.67259 \times 10^{-11}$ m^3/kg s^2, Isaac Newton's Gravitational Constant
$M_{uni}\, G = 1.49190478666 \times 10^{42}$ m^3/s^2

B) The Gravitational Structural Length of the Universe

$\mathcal{A}_{uni} = 1.6599679445 \times 10^{25}$ meters
$C^2 = 8.98755178737 \times 10^{16}$ m^2/s^2, velocity of light squared
$\mathcal{A}_{uni}\, C^2 = 1.49190478666 \times 10^{42}$ m^3/s^2

C) The Velocity of the Universe

$v_{uni} = 186{,}204{,}430.888$ m/s
$(v_{uni})^2 = 3.46720900823 \times 10^{16}$ m^2/s^2
$\mathcal{R}_{uni} = 4.30289833442 \times 10^{25}$ m
$(v_{uni})^2\, \mathcal{R}_{uni} = 1.49190478666 \times 10^{42}$ m^3/s^2

Therefore: $M_{uni}\, G = \mathcal{A}_{uni}\, C^2 = (v_{uni})^2\, \mathcal{R}_{uni}$

D) Three Energy Concepts of the Universe

1) First Energy Concept:

$E_{uni} = M_{uni}\, C^2$, Einstein's First Concept
$E_{uni} = 2.00950028879 \times 10^{69}$ Joules

Data:

$M_{uni} = 2.23587060896 \times 10^{52}$ kg, mass of the universe
$C = 2.99792458 \times 10^{16}$ m/s, velocity of light
$C^2 = 8.98755178737 \times 10^{16}$ m^2/s^2, velocity of light squared

2)   Second Energy Concept:

$E_{uni} = (\mathcal{A}_{uni} \, C^4) / G$

$E_{uni} = 2.0095028878 \times 10^{69}$ Joules

Data:

$P_c = C^4 / G = 1.21056571932 \times 10^{44}$ kg m/s^2

$\mathcal{A}_{uni} = 1.6599679445 \times 10^{25}$ meters, Gravitational Structural Length of the Universe

$C^4 = 8.07760871307 \times 10^{33}$ m^4/s^4

$G = 6.67259 \times 10^{-11}$ m^3/kg s^2, Isaac Newton's Gravitational Constant

Area Constant

$\mathcal{A}_{uni} = 1.6599679445 \times 10^{25}$ meters

$\lambda_{uni} = 9.88528079411 \times 10^{-95}$ m

$(\mathcal{A}_{uni} \times \lambda_{uni}) = 1.64092492406 \times 10^{-69}$ m^2

3)   Third Energy Concept

$E = \hbar f = \hbar C / \lambda = 2.00950028877 \times 10^{69}$ Joules

$f_{uni} = \dfrac{2.99792458 \times 10^8 \text{ m/s}}{9.88528079411 \times 10^{-95} \text{ m}}$

$f_{uni} = 3.03271565314 \times 10^{102}$ vibrations/s, frequency of the universe

$\hbar = 6.6260755 \times 10^{-34}$ kg m^2/ s, Planck's Constant

$E_{uni} = \hbar f_{uni} = 2.00950028877 \times 10^{69}$ Joules

Area Constant Check

$\mathbf{P} = (\hbar / C^3) = 2.45920238477 \times 10^{-59}$ kg s^2/ m, Pershing's Ratio Constant

Data:

$\hbar = 6.6260755 \times 10^{-34}$ kg m^2/ s, maximum Planck's constant
$C = 2.99792458 \times 10^{16}$ m/s, velocity of light
$C^3 = 2.69440024174 \times 10^{25}$ m^3/s^3
$(\hbar / C^3) = 2.45920238477 \times 10^{-59}$ kg s^2/ m
$G = 6.67259 \times 10^{-11}$ m^3/kg s^2, Isaac Newton's Gravitational Constant

Checks:

$(\mathcal{A} \times \lambda) = (\hbar / C^3)\, G = 1.64092492406 \times 10^{-69}$ m^2

The Density of the Universe

$$G_{uni} = \frac{M_{uni}}{(4/3)\,\pi\,(R_{uni})^3} = 6.7 \times 10^{-26} \text{ kg/m}^3$$

$$G_{uni} = \frac{2.23587060896 \times 10^{52} \text{ kg}}{(4/3)\,\pi\,(4.30289833442 \times 10^{25} \text{ m})^3}$$

$(4/3)\,\pi = 4.1887902048$

$(R_{uni})^3 = 7.9667878999 \times 10^{76} \text{ m}^3$

vibrations $_{uni} = (4/3)\,\pi\,(R_{uni})^3 = 3.33712031188 \times 10^{77} \text{ m}^3$

Final Conclusion

G_{uni} = density of the universe

$G_{uni} = 6.7 \times 10^{-26} \text{ kg/m}^3$

$$G_{uni} = \frac{M_{uni}}{(4/3)\,\pi\,(R_{uni})^3}$$

Aluminum

$\mathcal{A}_{\text{Aluminum}} = 3.32635908407 \times 10^{-53}$ meters, gravitational structural length of Aluminum

$\lambda_{\text{Aluminum}} = 4.93309616486 \times 10^{-17}$ meters, wavelength of Aluminum

$(\mathcal{A}_{\text{Aluminum}} * \lambda_{\text{Aluminum}}) = 1.64092492406 \times 10^{-69} \text{ m}^2$

Photoelectric Threshold
Frequency & Work Function of Cesium

* The threshold of frequency for cesium is in the visible range of the spectrum.

Work function of cesium $\Phi = 4.6 \times 10^{14}$ frequency

"The energy required to remove the surface electrons of cesium."

$\lambda_{blue} = 4.6 \times 10^{-7}$ m, wavelength of light

$l_{blue} = (C / \lambda_{blue}) = (2.99792458 \times 10^8 \text{ m/s}) / (4.6 \times 10^{-7} \text{ m})$

$l_{blue} = 6.51722734783 \times 10^{14}$ frequency

$E_{blue} = h\,l = 4.31836404574 \times 10^{-19}$ joules

$E_{blue} = \Phi = h\,l$

$KE = (E_{blue} - \Phi_{metal}) = (\frac{1}{2})(M_e)(v_e)^2 = 1.27036931574 \times 10^{-19}$ joules
$v_e = 528{,}123.332073$ m/s, velocity of electron from cesium
$(v_e)^2 = 278{,}914{,}253{,}880$ m^2/s^2
$M_e = 9.1093897 \times 10^{-31}$ kg, mass of electron

Energy of One Ampere Current

Mass flow of the electrons in the current:

A) Mass = (6.2415063631 x 10^{18} electrons/sec) x (9.1093897 x 10^{-31} kg/electron)

  = 1 ampere

B) Mass = 5.68563137765 x 10^{-12} kg

  E = Mass c^2 , Einstein's Energy Concept

  E = 510,999.064505 joules

Data:

c = 2.99792458 x 10^8 m/s

c^2 = 8.98755178737 x 10^{16} m^2/s^2

The Ampere – Coulomb Concept And Area Constant

A different approach

$$\text{One ampere} = \left[\frac{\text{One electron}}{1.60217733 \times 10^{-19}\text{coulomb}} \right]$$

$$\left[\frac{1}{\text{One ampere}} \right] = \left[\frac{6.24150636309 \times 10^{18} \text{ electrons}}{\text{One coulomb}} \right]$$

$$G \left[6.24150636309 \times 10^{18} \text{ electrons} \times \frac{9.1093897 \times 10^{-31} \text{ kg}}{1 \text{ electron}} \right]$$

$$G \left[5.6853137764 \times 10^{-12} \text{ kg} \right]$$

Isaac Newton's Gravitational Constant is $G = 6.67259 \times 10^{-11} \text{ m}^3/ \text{kg s}^2$

$$* \left[M_{amp} \right] \left[G \right] = 3.7978870741 \times 10^{-22} \text{ m}^3/ \text{s}^2$$

Gravitational Structural Length of the Coulomb
$$A_{amp} = 4.22115921795 \times 10^{-39} \text{ meters}$$

Velocity of light
$$C = 2.99792458 \times 10^{8} \text{ m/s}$$
$$C = 8.98755178737 \times 10^{16} \text{ m}^2/\text{s}^2$$
$$A_{amp} \, C^2 = 3.7978870741 \times 10^{-22} \text{ m}^3/ \text{s}^2$$

Therefore : $A_{amp} \, C^2 = M_{amp} \, G$

Three Energy Concepts

1) $E_{amp-coul} = M C^2$

   $E_{amp-coul} = 510,970.519928$ kg m^2/s$^2 \approx$ Joules

Data:

$M_{amp-coul} = 5.6853137764 \times 10^{-12}$ kg
$C = 2.99792458 \times 10^{16}$ m/s, velocity of light
$C^2 = 8.98755178737 \times 10^{16}$ m^2/s^2

2) $E_{amp-coul} = \hbar f_{amp-coul} = \hbar C / \lambda_{amp-coul}$

Data:

$\hbar = 6.62607.55 \times 10^{-34}$ Joules*s, maximum Planck's constant
$C = 2.99792458 \times 10^{16}$ m/s, velocity of light
$f_{amp} = 7.71151068122 \times 10^{38}$ vibrations/s, frequency of the ampere
$f = C / \lambda$
Therefore: $\lambda = C / f = 3.887373888 \times 10^{-31}$ m

3) $E_{amp-coul} = \dfrac{A_{coul} C^4}{G} = 510,970.519928$ kg m^2/s$^2 \approx$ Joules

   $E_{amp-coul} = \dfrac{A_{coul} C^4}{G} = 510,970.519928$ Joules

Data:

$A_{coul} = 4.22115921795 \times 10^{-39}$ meters, Gravitational Structural Length of the Coulomb
$A_{coul} C^2 = 3.79378870741 \times 10^{-22}$ m^3/s^2
$C^2 = 8.98755178737 \times 10^{16}$ m^2/s^2, velocity of light squared
$\dfrac{A_{coul} C^2}{G} = 5.68563137764 \times 10^{-12}$

- $G = 6.67259 \times 10^{-11}$ m^3/kg s^2, Isaac Newton's Gravitational Constant

Universal Area Constant

$\mathcal{A}_{amp} = 4.22116079244 \times 10^{-39}$ meters
$\lambda_{amp} = 3.8873783889 \times 10^{-31}$ meters
$(\mathcal{A}_{amp} \times \lambda_{amp}) = 1.64092492406 \times 10^{-69}$ m^2
$(\mathcal{A}_{amp} \times \lambda_{amp}) = (\hbar / C^3)\, G$

Pershing's Area Constant

$(\hbar / C^3)\, G = 1.64092492406 \times 10^{-69}$ m^2

Data:

$\hbar = 6.6260755 \times 10^{-34}$ kg m^2/ s, Planck's Constant
$C = 2.99792458 \times 10^{16}$ m/s, velocity of light
$C^3 = 2.69440024174 \times 10^{25}$ m^3/s^3
$\mathbf{P} = (\hbar / C^3) = 2.45920238477 \times 10^{-59}$ kg s^2/ m, Pershing's Ratio Constant
$G = 6.67259 \times 10^{-11}$ m^3/kg s^2, Isaac Newton's Gravitational Constant
$(\hbar / C^3)\, G = 1.64092492406 \times 10^{-69}$ m^2

Wavelength & Mass of an Electrical Current

Wavelength

$\text{E}_{\text{current}} = \hbar\, f_{\text{current}}$
$\text{E}_{\text{current}} = 3.31303775 \times 10^{-29}$ joules

$f_{\text{current}} = (C / \lambda_{\text{current}})$

$f_{\text{current}} = 50,000$ vibrations/s
$\lambda_{\text{current}} = 5995.84916$ meters
$C = 2.99792458 \times 10^{8}$ m/s

Mass

$\text{E}_{\text{current}} = M_{\text{current}}\, C^{2}$
$\text{E}_{\text{current}} = 3.31303775 \times 10^{-29}$ joules

$\text{M}_{\text{current}} = 3.68625163824 \times 10^{-46}$ kg
$C^{2} = 8.98755178737 \times 10^{16}$ m^2/s^2

Mass & Wavelength of the Proton

Mass of Proton = $\hbar / (C / \lambda_p)$
Mass of Proton = $1.6726231 \times 10^{-27}$ kg

$\hbar = 6.6260755 \times 10^{-34}$ joules s, Max Planck's constant
$C = 2.99792458 \times 10^{8}$ m/s, velocity of light
$\lambda_p = 1.32140999301 \times 10^{-15}$ m, wavelength of the proton

* Equation can also be changed to find the wavelength.

$\lambda_p = \hbar / (C / \text{Mass of Proton})$

The Proton Concept

1) The Gravitational Structural Length of the Proton:

$A_p = 2.81794096266 \times 10^{-15}$ m

$c^2 = 8.98755178737 \times 10^{16}$ m^2/s^2

$A_p\, c^2 = 253.263903357$ m^3/s^2

2) The velocity of the electron in the 1st shell of the Hydrogen atom:

$v_e = 2187691.43195$ m/s

$(v_e)^2 = 4.78599380143 \times 10^{12}$ m^2/s^2

$a_o = 5.29177249 \times 10^{-11}$ m, Bohr's radius

$(v_e)^2\, (a_o) = 253.263903357$ m^3/s^2

3) $A_p\, c^2 = (v_e)^2\, (a_o) = 253.263903357$ m^3/s^2

Three Energy Concepts of the Proton

*First Energy Concept
    Albert Einstein's Energy Concept

$$E_p = M_p C^2$$
$$E_p = 1.5032786732 \times 10^{-10} \text{ joules}$$

Data:
    $M_p = 1.6726231 \times 10^{-27}$ kg, mass of proton
    $C = 2.99792458 \times 10^8$ m/s, velocity of light
    $C^2 = 8.98755178737 \times 10^{16}$ m^2/s^2

*Second Energy Concept
    Max Planck's Energy Concept

$$E_p = \hbar f_p = \hbar (C / \lambda_p) = 1.5032786732 \times 10^{-10} \text{ joules}$$

Data:
    $\hbar = 6.6260755 \times 10^{-34}$ joules s, Max Planck's constant
    $f_p = 2.26873157904 \times 10^{23}$ vib/s, frequency of proton
    $C = 2.99792458 \times 10^8$ m/s, velocity of light
    $\lambda_p = 1.32140999301 \times 10^{-15}$ m, wavelength of the proton
    $f_p = (C / \lambda_p) = 2.26873157904 \times 10^{23}$ vib/s

*Third Energy Concept
    Don J. Pershing's Energy Concept

$$E_p = (\mathcal{A}_p C^4) / G_P$$
$$E_p = \mathcal{A}_p \mathbf{P}_{c_1}$$
$$E_p = 1.5032786732 \times 10^{-10} \text{ joules}$$

Data:
    $\mathcal{A}_p = 2.81794096266 \times 10^{-15}$ m, Gravitational Structural Length of the proton
    $C = 2.99792458 \times 10^8$ m/s, velocity of light
    $C^4 = 8.07760871306 \times 10^{33}$ m^4/s^4
    $\mathbf{P}_{c_1} = (\hbar / C^3) = 2.45920238477 \times 10^{-59}$ kg s^2/m, Pershing's First Constant
    $G_P = 1.51417198146 \times 10^{29}$ m^3/ kg s^2 ,
    Pershing's Gravitational Constant of the Hydrogen Atom
    $C^3 = 2.69440024174 \times 10^{25}$ m^3/s^3
    $\mathcal{A}_p C^4 = 2.2762244729 \times 10^{19}$ m^5/s^4

*Fourth Concept
    Pershing's 2nd Concept

$E_p = (\mathcal{A}_p C^2) Z_P = 1.5032786732 \times 10^{-10}$ joules

Data:

$(\mathcal{A}_p C^2) = 253.263903357$ m^3/s^2
$\mathcal{A}_p = 2.81794096266 \times 10^{-15}$ m, Gravitational Structural Length of the proton
$C = 2.99792458 \times 10^8$ m/s, velocity of light
$C^2 = 8.98755178737 \times 10^{16}$ m^2/s^2

$Z_P = \dfrac{C^2}{G_P} = \dfrac{(8.98755178737 \times 10^{16} \text{ m}^2/\text{s}^2)}{(1.51417198146 \times 10^{29} \text{ m}^3/\text{kg s}^2)}$
$Z_P = 5.93562151289 \times 10^{-13}$ kg/m
$G_P = 1.51417198146 \times 10^{29}$ m^3/ kg s^2,
Pershing's Gravitational Constant of the Hydrogen Atom

$E_p = (\mathcal{A}_p C^2) Z_P = 1.5032786732 \times 10^{-10}$ joules

Compton's Wavelength of the Proton and Electron

First Concept: Wavelength of the Electron

$$\lambda_e = \frac{h}{M_e \, C} = 2.42631060001 \times 10^{-12} \text{ m}$$

Data:

$$P_c = \frac{h}{C} = \frac{6.6260755 \times 10^{-34} \text{kg m}^2/\text{s}}{2.99792458 \times 10^8 \text{ m/s}}$$

$$P_c = \frac{h}{C} = 2.21022087887 \times 10^{-42} \text{ kg m}$$

$$M_e = 9.1093897 \times 10^{-31} \text{ kg, mass of electron}$$

$$\lambda_e = \left(\frac{h}{C}\right)\left(\frac{1}{M_e}\right)$$

$$\lambda_e = 2.42631060001 \times 10^{-12} \text{ m}$$

Second Concept: Wavelength of the Proton

$$\lambda_p = \frac{h}{M_p \, C} = 1.321409993 \times 10^{-15} \text{ m}$$

Data:

$$P_c = \frac{h}{C} = 2.21022087887 \times 10^{-42} \text{ kg m}$$

$$M_p = 1.6726231 \times 10^{-27} \text{ kg, mass of the proton}$$

$$\lambda_p = \frac{h}{M_p \, C} = 1.321409993 \times 10^{-15} \text{ m}$$

Ratio Concept:

$$\frac{\lambda_e}{\lambda_p} = \frac{2.42631060001 \times 10^{-12}\text{m}}{1.321409993 \times 10^{-15}\text{ m}}$$

$$\frac{\lambda_e}{\lambda_p} = 1836.15275566$$

Therefore:

$$\frac{\lambda_e}{\lambda_p} = \frac{M_p}{M_e} = \frac{\mathcal{A}_p}{\mathcal{A}_e} = 1836.15275566$$

The Velocity of the Electron

$(v_e)^2 (a_o) = A_p C^2$

Therefore: $(v_e) = C \sqrt{A_p / (a_o)}$

Data:

$A_p =$ 2.81794096266 x 10^{-15} m = 5.32513627142 x 10^{-5} radians

(a_o) 5.29177249 x 10^{-11} m

$\sqrt{A_p / (a_o)}$ = 7.2973531307 x 10^{-3} radians

C = 2.99792458 x 10^8 m/s, velocity of light

$(v_e) = C \sqrt{A_p / (a_o)}$ = 2,187,691.43195 m/s
average orbital velocity of the electron in 1st shell of the hydrogen atom

$(v_e)^2$ = 4.78599380143 x 10^{12} m²/s²

$(v_e)^2 (a_o)$ = 253.263903357 m³/s²

Max Planck's Concept
of
The Velocity of the Electron

$v_e = (\hbar/2\pi)\,(1/M_e)\,(1/\,a_o)$

$v_e = 2{,}187{,}691.43195$ m/s, velocity of the electron

$(v_e)^2 = 4.78599380143 \times 10^{12}$ m^2/s^2

$(v_e)^2\,(a_o) = 253.263903357$ m^3/s^2

Data:

    $\hbar = 6.6260755 \times 10^{-34}$ m^2/s, Max Planck's Constant

    $2\pi = 6.28318530718$

    $(\hbar/2\pi) = 1.05457266913 \times 10^{-34}$ kg m^2/s

    $M_e = 9.1093897 \times 10^{-31}$ kg, mass of electron

    $(\hbar/2\pi)(1/M_e) = 1.15767653362 \times 10^{-4}$ m^2/s

    $a_o = 5.29177249 \times 10^{-11}$ m, Bohr's radius

$v_e = (\hbar/2\pi)\,(1/M_e)\,(1/\,a_o) = 2{,}187{,}691.43195$ m/s, the average orbital velocity of the
electron in the 1st shell of the hydrogen atom

$(v_e)^2 = 4.78599380143 \times 10^{12}$ m^2/s^2

$(v_e)^2\,(a_o) = 253.263903357$ m^3/s^2

Max Planck's Method:
Calculation of the Velocity of the Electron

$v_e = (\hbar/2\pi)\,(1/M_e)\,(1/a_o)$, Max Planck's Method

$v_e = 2{,}187{,}691.43195$ m/s *this is the velocity of the electron in the 1st shell of the Hydrogen atom

Data:

$\hbar = 6.6260755 \times 10^{-34}$ m^2/s, Max Planck's Constant

$2\pi = 6.28318530718$

$\hbar = (\hbar/2\pi) = 1.05457266913 \times 10^{-34}$ kg m^2/s   *$\hbar$ is called h-bar

$M_e = 9.1093897 \times 10^{-31}$ kg, mass of electron

$\hbar(1/M_e) = 1.15767653362 \times 10^{-4}$ m^2/s

$a_o = 5.29177249 \times 10^{-11}$ m, Bohr's radius

$(v_e)^2 = 4.78599380143 \times 10^{12}$ m^2/s^2

$(v_e)^2\,(a_o) = 253.263903357$ m^3/s^2

*The Kinetic Concept

$KE_e = \frac{1}{2}\,(M_e)\,(v_e)^2$, Planck's Concept

$KE_e = 2.17987413195 \times 10^{-18}$ joules

$$\frac{2\,(KE_e)(a_o)}{M_e} = 253.263903357 \text{ m}^3/\text{s}^2$$

Data:

$2\,(KE_e) = 4.3597482639 \times 10^{-18}$ joules

$a_o = 5.29177249 \times 10^{-11}$ m, Bohr's radius

$2\,(KE_e)(a_o) = 2.30707959262 \times 10^{-28}$ joules m

$M_e = 9.1093897 \times 10^{-31}$ kg, mass of electron

Final Conclusion:

$$A_p\,C^2 = (v_e)^2\,(a_o) = \frac{2\,(KE_e)(a_o)}{M_e} = M_p\,G_P = [(\hbar/2\pi)\,(1/M_e)\,(1/a_o)]^2\,(a_o)$$

$= 253.263903357$ m^3/s^2

Pershing's Method for the
Velocity of the Electron

1) $(v_e)^2 (a_o) = A_p \, C^2 = 253.263903357 \ m^3/s^2$

2) $M_p \, G_P = 253.263903357 \ m^3/s^2$

Data:

$M_p = 1.6726231 \times 10^{-27}$ kg, mass of proton

$G_P = 1.51417198146 \times 10^{29} \ m^3/kg \ s^2,$

Pershing's *Gravitational* Constant of the Proton

Note:

$G = 6.67259 \times 10^{-11} \ m^3/kg \ s^2$, Newton's Gravitational Constant

$N_p = 2.26924175089 \times 10^{39}$

$G_P = (G)(N_p) = 1.51417198146 \times 10^{29} \ m^3/kg \ s^2$

$$G_P = \frac{A_p \, C^2}{M_p} = 1.51417198146 \times 10^{29} \ m^3/kg \ s^2$$

Pershing's Inverse Fine Structure Method
for the Velocity of the Electron

$$(v_e) = \frac{C}{\propto^{-1}} = \frac{C}{\text{Inverse Fine Structure Constant}}$$

$(v_e) = 2{,}187{,}691.43195$ m/s

Data:

$C = 2.99792458 \times 10^8$ m/s, velocity of light

$\propto^{-1} = 137.035988541$, Inverse Fine Structure Constant

$\sqrt{A_p / (a_o)} = 7.2973531307 \times 10^{-3}$ radians, Fine Structure Constant

The inverse of $\sqrt{A_p / (a_o)}$ is equal to $\propto^{-1}$

Three Concepts for the Kinetic Energy of the Electron

First Concept

$KE_e = \frac{1}{2} (M_e) (v_e)^2$, Planck's Concept

$KE_e = 2.17987413195 \times 10^{-18}$ joules

Data:

$(M_e) = 9.1093897 \times 10^{-31}$ kg, mass of electron

$(v_e) = 2,187,691.43195$ m/s, velocity of electron in 1st shell of Hydrogen atom

$(v_e)^2 = 4.78599380143 \times 10^{12}$ m^2/s^2

$(M_e) (v_e)^2 = 4.3597482639 \times 10^{-18}$ joules

$KE_e = \frac{1}{2} (M_e) (v_e)^2 = 2.17987413195 \times 10^{-18}$ joules

Second Concept

$KE_e = \frac{1}{2} (M_e) [(\mathcal{A}_p\, C^2) / a_o]$, Planck's Concept

$KE_e = 2.17987413195 \times 10^{-18}$ joules

Data:

$\mathcal{A}_p\, C^2 = 253.263903357$ m^3/s^2

$\mathcal{A}_p = 2.81794096266 \times 10^{-15}$ m,

Pershing's Gravitational Structural Length of the proton

$C = 2.99792458 \times 10^8$ m/s, velocity of light

$C^2 = 8.98755178737 \times 10^{16}$ m^2/s^2

$[(\mathcal{A}_p\, C^2) / a_o] = 4.78599380143 \times 10^{12}$ m^2/s^2

$a_o = 5.29177249 \times 10^{-11}$ m, Bohr's radius

$(M_e) = 9.1093897 \times 10^{-31}$ kg, mass of electron

$KE_e = \frac{1}{2} (M_e) [(\mathcal{A}_p\, C^2) / a_o] = 2.17987413195 \times 10^{-18}$ joules

Third Concept

$KE_e = (\hbar) (f_e) = (\hbar) (C / \lambda_e)$

$KE_e = 2.17987413195 \times 10^{-18}$ joules

$\hbar = 6.6260755 \times 10^{-34}$ m^2/s, Max Planck's Constant

$C = 2.99792458 \times 10^8$ m/s, velocity of light

$f_e = (C / \lambda_e) = 3.28984197652 \times 10^{15}$ vibrations /s

$\lambda_e = 9.11267046076 \times 10^{-8}$ m,

wavelength of the electron – "Pershing's wavelength value"

Ratio Concepts of the Proton and Electron

$$\frac{M_p}{M_e} = \frac{1.6726231 \times 10^{-27} \text{ kg}}{9.1093897 \times 10^{-31} \text{ kg}} = 1{,}836.15275566$$

$$\frac{\mathcal{A}_p}{\mathcal{A}_e} = \frac{2.81794096266 \times 10^{-15} \text{ m}}{1.53469854508 \times 10^{-18} \text{ m}} = 1{,}836.15275566$$

$$\mathcal{A}_e \, C^2 = 0.137931826519 \text{ m}^3/\text{s}^2$$
$$C^2 = 8.98755178737 \times 10^{16} \text{ m}^2/\text{s}^2$$

$$G_P = (\mathcal{A}_e \, C^2) \, / \, M_e = 0.137931826549 \text{ m}^3/\text{s}^2$$
$$G_P = 1.51417198146 \times 10^{-29} \text{ m}^3/\text{kg s}$$

$$\mathcal{A}_p \, C^2 = 253.263903357 \text{ m}^3/\text{s}^2$$
$$\mathcal{A}_p = 2.81794096266 \times 10^{-15} \text{ m, Gravitational Structural Length of the Proton}$$
$$C^2 = 8.98755178737 \times 10^{16} \text{ m}^2/\text{s}^2$$

$$(\mathcal{A}_p \, C^2) \, / \, M_p = 1.51417198146 \times 10^{29} \text{ m}^3/\text{kg s}^2 = G_P$$

Ratios of Gravitational Structural Lengths & Masses

1) $\dfrac{\mathcal{A}_p}{\mathcal{A}_e} = \dfrac{2.81794096266 \times 10^{-15} \text{ m}}{1.53469854508 \times 10^{-18} \text{ m}}$

$\dfrac{\mathcal{A}_p}{\mathcal{A}_e} = 1{,}836.15275566$

2) $\dfrac{\mathcal{A}_d}{\mathcal{A}_e} = \dfrac{5.63308491407 \times 10^{-15} \text{ m}}{1.53469854508 \times 10^{-18} \text{ m}}$

$\dfrac{\mathcal{A}_d}{\mathcal{A}_e} = 3{,}670.48299624$

3) $\dfrac{\mathcal{A}_\rho}{\mathcal{A}_e} = \dfrac{3.17326954206 \times 10^{-16} \text{ m}}{1.53469854508 \times 10^{-18} \text{ m}}$

$\dfrac{\mathcal{A}_\rho}{\mathcal{A}_e} = 206.76826418$

4) $\dfrac{\mathcal{A}_d}{\mathcal{A}_p} = \dfrac{5.63308491407 \times 10^{-15} \text{ m}}{2.81794096266 \times 10^{-15} \text{ m}}$

$\dfrac{\mathcal{A}_d}{\mathcal{A}_p} = 1.99900742732$

Note: $\mathcal{A}_p \, C^2 = v^2 \, a$ concept will replace Max Planck's concept of $v = \left[\dfrac{\hbar}{2\pi}\right]\left[\dfrac{1}{M_e}\right]\left[\dfrac{1}{a}\right]$

with $v_e = C \sqrt{(\mathcal{A}_p / a)}$; my method is much faster.

Ratios of Masses

1) $\dfrac{M_p}{M_e} = \dfrac{1.6726231 \times 10^{-27} \text{ kg}}{9.1093897 \times 10^{-31} \text{ kg}}$

$\dfrac{M_p}{M_e} = 1{,}836.15275566$

2) $\dfrac{M_d}{M_e} = \dfrac{3.3435860 \times 10^{-27} \text{ kg}}{9.1093897 \times 10^{-31} \text{ kg}}$

$\dfrac{M_d}{M_e} = 3{,}670.48299624$

3) $\dfrac{M_\rho}{M_e} = \dfrac{1.8835327 \times 10^{-28} \text{ kg}}{9.1093897 \times 10^{-31} \text{ kg}}$

$\dfrac{M_\rho}{M_e} = 206.76826418$

4) $\dfrac{M_d}{M_p} = \dfrac{3.3435860 \times 10^{-27} \text{ kg}}{1.6726231 \times 10^{-27} \text{ kg}}$

$\dfrac{M_d}{M_p} = 1.99900742732$

Note: $\dfrac{\mathcal{A}_p}{\mathcal{A}_e} = 1{,}836.15275566 = \dfrac{M_p}{M_e}$

* Therefore the ratios of masses is the same as the ratios of gravitational structural lengths. This is always true.

* Key Concept:

It is much faster using Gravitational Structural Length Concepts. Example below:

A) $\mathcal{A}_p = 2.81794096266 \times 10^{-15}$ m
       $C^2 = 8.98755178737 \times 10^{16}$ m^2/s^2
       $\mathcal{A}_p\,C^2 = 253.263903357$ m^3/s^2

B) $\mathcal{A}_e = 1.53469854508 \times 10^{-18}$ m
       $\mathcal{A}_e\,C^2 = 0.137931826519$ m^3/s^2

Neutrons, Muons, & Deuterons

- ## Neutrons

1) Mass of neutron = $1.6749286 \times 10^{-27}$ kg
   $G_P = 1.51417198146 \times 10^{29}$ m^3/ kg s^2 ,
   Pershing's Gravitational Constant of the Hydrogen Atom

   $(M_{neutron})\,(G_P) = 253.612995707$ m^3/s^2

2) $\mathcal{A}_{neutron} = 2.82182513899 \times 10^{-15}$ m
   $C^2 = 8.98755178737 \times 10^{16}$ m^2/s^2

   $\mathcal{A}_{neutron}\, C^2 = 253.612995707$ m^3/s^2

- ## Muons

1) Mass of muon   = $1.88353270 \times 10^{-28}$ kg
   $G_P = 1.51417198146 \times 10^{29}$ m^3/ kg s^2

   $(M_{muon})\,(G_P) = 28.519924405$ m^3/s^2

2) $\mathcal{A}_{muon} = 3.17326954878 \times 10^{-16}$ m
   $C^2 = 8.98755178737 \times 10^{16}$ m^2/s^2

   $\mathcal{A}_{muon}\, C^2 = 28.5199244049$ m^3/s^2

- ## Deuteron

1) Mass of the deuteron = $3.3435860 \times 10^{-27}$ kg
   $G_P = 1.51417198146 \times 10^{29}$ m^3/ kg s^2

   $(M_{deuteron})\,(G_P) = 506.27642388$ m^3/s^2

2) $\mathcal{A}_{deuteron} = 5.63308491407 \times 10^{-15}$ m
   $C^2 = 8.98755178737 \times 10^{16}$ m^2/s^2

   $\mathcal{A}_{deutron}\, C^2 = 506.276423879$ m^3/s^2

Balmer Series – The Visible Line Spectrum

Note: A visible line spectrum or wavelength is emitted when an electron at the third level of the hydrogen atom jumps back to the second shell. Only electrons returning to the second produce a visible line spectra or wavelength.

Balmer's Line Spectrum Equation

$$\lambda_H = \frac{1}{R_\infty \left(1/n^2 - 1/m^2 \right)}$$

$\lambda_H = 6.56112273176 \times 10^{-7}$ meters wavelength emitted

$R_\infty = 10{,}973{,}731.62241$ meters, Rydberg Constant

$n = 2$ for 2nd shell & $n^2 = 4$
$m = 3$ for 3rd shell & $m^2 = 9$

$1/4 - 1/9 = (0.25 - 0.11111...)$
$1/4 - 1/9 = 0.138888...$

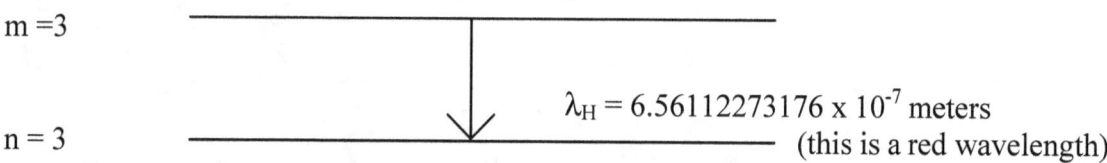

m = 3

n = 3

$\lambda_H = 6.56112273176 \times 10^{-7}$ meters
(this is a red wavelength)

$$\lambda_H = 6.56112273176 \times 10^{-7} \text{ m} \times \frac{A^\circ}{10^{10} \text{ m}}$$

$\lambda_H = 6561.12273176 \text{ A}^\circ$, Red Wavelength

Note: A° = Angstrom

Wavelength of Light

* Basic Equation:

$$\lambda = \frac{Y\,d}{m\mathcal{R}} = \qquad \text{where } m = 3^{\text{rd}} \text{ bright fringe formed}$$

$$\lambda = \frac{(7.8 \times 10^{-3}\ \text{m})\,(2 \times 10^{-4}\ \text{m})}{(3)\,(1\ \text{m})} = 5.2 \times 10^{-7}\ \text{m} = 520\ \text{nm, green light wavelength}$$

$$E_E = \hbar f_E = \frac{\hbar C}{\lambda} \qquad 1^{\text{st}} \text{ Concept}$$

$\hbar = 6.6260755 \times 10^{-34}$ Joules * s

$C = 2.99792458 \times 10^{8}$ m/s

$\lambda = 5.2 \times 10^{-7}$ m

$$E_E = \hbar f_E = 3.82009127123 \times 10^{19} \text{ Joules, Planck's Energy Concept}$$

Properties of Light

Concept of $(A)(C^2) = (M)(G)$

1) $(A_{red})(C^2) = (M_{red})(G) = 2.30435902096 \times 10^{-46} \ m^3/s$
2) $(A_{orange})(C^2) = (M_{orange})(G) = 2.49964368375 \times 10^{-46} \ m^3/s$
3) $(A_{yellow})(C^2) = (M_{yellow})(G) = 2.58735012548 \times 10^{-46} \ m^3/s$
4) $(A_{green})(C^2) = (M_{green})(G) = 2.50836444402 \times 10^{-46} \ m^3/s$
5) $(A_{blue})(C^2) = (M_{blue})(G) = 3.13785058173 \times 10^{-46} \ m^3/s$
6) $(A_{violet})(C^2) = (M_{violet})(G) = 3.20606472482 \times 10^{-46} \ m^3/s$

$C = 2.99792458 \times 10^8 \ m/s$
$C^2 = 8.98755178737 \times 10^{16} \ m^2/s^2$
$G = 6.67259 \times 10^{-11} \ m^3/kg \ s^2$, Isaac Newton's Gravitational Constant

Gravitational Structural Lengths of Light

1) $A_{red} = 2.64665310332 \times 10^{-63} \ m$
2) $A_{orange} = 2.82918090355 \times 10^{-63} \ m$
3) $A_{yellow} = 3.03874985937 \times 10^{-63} \ m$
4) $A_{green} = 3.28184984812 \times 10^{-63} \ m$
5) $A_{blue} = 3.56722809578 \times 10^{-63} \ m$
6) $A_{violet} = 3.9069641049 \times 10^{-63} \ m$

* $(\lambda_{red} \ x \ A_{red}) = 1.64092492406 \times 10^{-69} \ m^2$
* $(\lambda_{blue} \ x \ A_{blue}) = 1.64092492406 \times 10^{-69} \ m^2$

Mass of Light

1) $M_{red} = 3.56487238528 \times 10^{-36} \ kg$
2) $M_{orange} = 3.81072565323 \times 10^{-36} \ kg$
3) $M_{yellow} = 4.09300162755 \times 10^{-36} \ kg$
4) $M_{green} = 4.42044175776 \times 10^{-36} \ kg$
5) $M_{blue} = 4.80482799755 \times 10^{-36} \ kg$
6) $M_{violet} = 5.26243066399 \times 10^{-36} \ kg$

* use the gravitational and velocity of light constants to interchange between Mass and Gravitational Structural Lengths of Light
$M = \dfrac{A C^2}{G}$, where $C^2 = 8.98755178737 \times 10^{16} \ m^3/s^2$

The Energy of Light

Einstein's Energy Concept

1) $E_{red} = (M_{red})(C^2) = (A_{red})(C^4/G) = 3.20394751781 \times 10^{-19}$ joules
2) $E_{orange} = (M_{orange})(C^2) = (A_{orange})(C^4/G) = 3.42490941559 \times 10^{-19}$ joules
3) $E_{yellow} = (M_{yellow})(C^2) = (A_{yellow})(C^4/G) = 3.67860640934 \times 10^{-19}$ joules
4) $E_{green} = (M_{green})(C^2) = (A_{green})(C^4/G) = 3.97289492209 \times 10^{-19}$ joules
5) $E_{blue} = (M_{blue})(C^2) = (A_{blue})(C^4/G) = 4.31836404569 \times 10^{-19}$ joules
6) $E_{violet} = (M_{violet})(C^2) = (A_{violet})(C^4/G) = 4.72963681201 \times 10^{-19}$ joules

$C^2 = 8.98755178737 \times 10^{16}$ m^2/s^2
$(C^4/G) = \mathbf{P}_{c_2} = 1.21056571932 \times 10^{44}$ kg m/s^2

Max Planck's Energy Concept

1) $E_{red} = [(\hbar C)/\lambda_{red}] = 3.20394751781 \times 10^{-19}$ joules
2) $E_{orange} = [(\hbar C)/\lambda_{orange}] = 3.42490941559 \times 10^{-19}$ joules
3) $E_{yellow} = [(\hbar C)/\lambda_{yellow}] = 3.67860640934 \times 10^{-19}$ joules
4) $E_{green} = [(\hbar C)/\lambda_{green}] = 3.97289492209 \times 10^{-19}$ joules
5) $E_{blue} = [(\hbar C)/\lambda_{blue}] = 4.31836404574 \times 10^{-19}$ joules
6) $E_{violet} = [(\hbar C)/\lambda_{violet}] = 4.72963681201 \times 10^{-19}$ joules

$\hbar C = 1.98644746104 \times 10^{-25}$ joules m
$\hbar = 6.6260755 \times 10^{-34}$ joules s
$C = 2.99792458 \times 10^{8}$ m/s

Pershing's Energy Concept

1) $E_{red} = [\,((A_{red} C^2)/G)*(C^2)\,] = 3.20394751781 \times 10^{-19}$ joules
2) $E_{orange} = [\,((A_{orange} C^2)/G)*(C^2)\,] = 3.42490941559 \times 10^{-19}$ joules
3) $E_{yellow} = [\,((A_{yellow} C^2)/G)*(C^2)\,] = 3.67860640934 \times 10^{-19}$ joules
4) $E_{green} = [\,((A_{green} C^2)/G)*(C^2)\,] = 3.97289492209 \times 10^{-19}$ joules
5) $E_{blue} = [\,((A_{blue} C^2)/G)*(C^2)\,] = 4.31836404574 \times 10^{-19}$ joules
6) $E_{violet} = [\,((A_{violet} C^2)/G)*(C^2)\,] = 4.72963681201 \times 10^{-19}$ joules

Reference Sources

- The VNR Concise Encyclopedia of Mathematics

  By Von Nostrand Reinhold Company

- Exploring the Cosmos 2nd Edition

  By Louise Bermon & J.C. Evans

- The Mechanical Universe

  By Richard Vlenick, Tom Apostal, & David Goodstein

  Cambridge University Press

- Mathematical Principles of Natural Philosophy

  By Sir Isaac Newton

  Encyclopedia Britannica, Inc

- The Planets - Their Origin & Development

  By Harold C. Urey

  Oxford University Press, 1952

- Space Facts – Handbook of Basic & Advance Space Flight and Environmental Data for Scientist Engineers

  By General Electric

- Handbook of Chemistry & Physics 79th Edition

  David R. Lide, Editor-in-Chief

- Exploration of the Universe

  By George Abell

  University of California, LA

- <u>Introductory Astronomy</u> 2nd Edition

  By Nicholas Panonides & Thomas Arny

  Addison-Wesley Publishing Company

- <u>Einstein's Legacy - The Unity of Space & Time</u>

  By Julian Schwinger

  Scientific American Books, Inc.

- <u>Adventures In Celestial Mechanics – A First Course in the Theory of Orbits</u>

  By Victor Szebehely

- <u>Einstein's Theory of Relativity</u> Rev. Edition

  By Max Born

  Dover Publications, Inc.

- <u>Atomic Physics</u> 8th Edition

  By Max Born

  Dover Publications, Inc.

- Calculater Used: Hewlett Packard 48 G+

About the Author

Mr. Donald Pershing became interested in Astronomy and Physics as a young child while watching his father teach quantum physics and astronomy at the University of Indiana. His father would take him and his younger brother to look at the moon and the stars through the telescope at the University. Seeing the moon left Mr. Pershing wondering about the other planets in the Universe. This curiosity resulted in the building of a telescope by Mr. Pershing's brother while attending college.

While Mr. Pershing's father was earning his doctorate's degree at the University of Chicago, he would occasionally bring young Mr. Pershing to the school. On one such instance, Mr. Pershing attended a summer picnic with his father, along with many notable scientists. Mr. Pershing fondly remembers sitting on Dr. Albert Einstein's lap and being fed watermelon. This had a significant influence on the course of Mr. Pershing's career and fueled his passion for the study of the Universe.